Chemistry
DeMYSTiFieD®

DeMYSTiFieD® Series

Accounting Demystified
Advanced Calculus Demystified
Advanced Physics Demystified
Advanced Statistics Demystified
Algebra Demystified
Alternative Energy Demystified
Anatomy Demystified
Astronomy Demystified
Audio Demystified
Biology Demystified
Biophysics Demystified
Biotechnology Demystified
Business Calculus Demystified
Business Math Demystified
Business Statistics Demystified
C++ Demystified
Calculus Demystified
Chemistry Demystified
Circuit Analysis Demystified
College Algebra Demystified
Corporate Finance Demystified
Data Structures Demystified
Databases Demystified
Differential Equations Demystified
Digital Electronics Demystified
Earth Science Demystified
Electricity Demystified
Electronics Demystified
Engineering Statistics Demystified
Environmental Science Demystified
Everyday Math Demystified
Fertility Demystified
Financial Planning Demystified
Forensics Demystified
French Demystified
Genetics Demystified
Geometry Demystified
German Demystified
Home Networking Demystified
Investing Demystified
Italian Demystified
Java Demystified
JavaScript Demystified
Lean Six Sigma Demystified
Linear Algebra Demystified

Logic Demystified
Macroeconomics Demystified
Management Accounting Demystified
Math Proofs Demystified
Math Word Problems Demystified
MATLAB® Demystified
Medical Billing and Coding Demystified
Medical Terminology Demystified
Meteorology Demystified
Microbiology Demystified
Microeconomics Demystified
Nanotechnology Demystified
Nurse Management Demystified
OOP Demystified
Options Demystified
Organic Chemistry Demystified
Personal Computing Demystified
Pharmacology Demystified
Philosophy Demystified
Physics Demystified
Physiology Demystified
Pre-Algebra Demystified
Precalculus Demystified
Probability Demystified
Project Management Demystified
Psychology Demystified
Quality Management Demystified
Quantum Mechanics Demystified
Real Estate Math Demystified
Relativity Demystified
Robotics Demystified
Sales Management Demystified
Signals and Systems Demystified
Six Sigma Demystified
Spanish Demystified
SQL Demystified
Statics and Dynamics Demystified
Statistics Demystified
Technical Analysis Demystified
Technical Math Demystified
Trigonometry Demystified
UML Demystified
Visual Basic 2005 Demystified
Visual C# 2005 Demystified
XML Demystified

Chemistry
DeMYSTiFieD®

Linda D. Williams

Second Edition

McGraw Hill

New York   Chicago   San Francisco   Lisbon   London   Madrid   Mexico City
Milan   New Delhi   San Juan   Seoul   Singapore   Sydney   Toronto

The **McGraw·Hill** Companies

Cataloging-in-Publication Data is on file with the Library of Congress

McGraw-Hill books are available at special quantity discounts to use as premiums and sales promotions, or for use in corporate training programs. To contact a representative, please e-mail us at bulksales@mcgraw-hill.com.

Chemistry DeMYSTiFieD®, Second Edition

1 2 3 4 5 6 7 8 9 0   DOC/DOC   1 7 6 5 4 3 2 1

ISBN 978-0-07-175130-8
MHID 0-07-175130-0

Sponsoring Editor Judy Bass	**Project Manager** Vasundhara Sawhney, Glyph International	**Indexer** Robert A. Saigh
Editorial Supervisor Stephen M. Smith	**Copy Editor** Priyanka Sinha, Glyph International	**Cover Illustration** Lance Lekander
Production Supervisor Richard C. Ruzycka		**Art Director, Cover** Jeff Weeks
Acquisitions Coordinator Michael Mulcahy	**Proofreader** Erica Orloff	**Composition** Glyph International

Contents

	Preface	*xi*
	Acknowledgments	*xiii*
CHAPTER 1	**Chemistry**	**1**
	What Is Matter?	2
	What Is Modern Chemistry?	2
	Basic and Applied Science	4
	Scientific Method	6
	Hypothesis	7
	Measurements	8
	Precision versus Accuracy	13
	Conversion Factors	15
	Quiz	18
CHAPTER 2	**Atomic Structure and Theory**	**21**
	What Are Atoms?	22
	Beginnings of Atomic Theory	22
	Molecules	26
	Quiz	30
CHAPTER 3	**Elements and the Periodic Table**	**33**
	What Is Matter?	34
	Chemical Nomenclature	35
	Atomic Number	36
	Atomic Weight	37
	Classes of Elements	41
	Periods and Groups	42
	Metallurgy—The Chemistry of Metals	43
	Quiz	53

CHAPTER 4	**Solids and Liquids**	**55**
	What Are Solids?	56
	Crystallization and Bonding	57
	Properties of a Solid	60
	Mixtures	63
	Compounds	64
	What Are Liquids?	64
	Quiz	74
CHAPTER 5	**Gases and the Gas Laws**	**77**
	What Are Gases?	78
	Atmosphere	79
	Kinetic Energy and Gas Theory	80
	Atmospheric Pressure	81
	Empirical Gas Laws	83
	Avogadro's Law	87
	Ideal Gas Law	88
	Dalton's Law of Partial Pressures	90
	Quiz	92
CHAPTER 6	**Solutions**	**95**
	What Is a Solution?	96
	Solubility Rules	96
	What Is a Mole?	98
	Molar Mass	98
	Molarity	100
	Percent Solution	100
	Changing the Concentration	103
	Quiz	106
CHAPTER 7	**Orbitals**	**109**
	What Are Orbitals?	110
	Electron Energy Levels	110
	Subshells and the Periodic Table	115
	Ionization Energy	118
	Valence Bond Theory	119
	Molecular Orbital Theory	120
	Resonance Theory	121
	Molecular Geometry	123
	Quiz	125
CHAPTER 8	**Chemical Bonds**	**127**
	What Are Covalent Bonds?	128

Covalent Compounds 130
Polarity 131
Naming Covalent Compounds 133
What Are Ions? 134
Ionic Bonds 138
Quiz 141

CHAPTER 9   Electrochemistry   143
Introduction to Electrochemistry 144
What Is Oxidation and Reduction (Redox)? 144
Balancing Redox Reactions 145
Oxidation State 148
What Is an Electrochemical Cell? 149
What Is Electrolysis? 153
Conductors and Insulators 154
Quiz 156

CHAPTER 10   Acids and Bases   159
What Are Acids and Bases? 160
Arrhenius Theory 160
Brønsted-Lowry Acids and Bases 161
Neutralization 162
Conjugate Acid-Base Pairs 162
Why Is Hydrogen Important? 166
pH Scale 167
Buffers 169
Acids, Bases, and Safety 170
Quiz 172

CHAPTER 11   Thermodynamics   175
What Is Thermodynamics? 176
Potential and Kinetic Energy 176
What Is Standard State? 176
First Law of Thermodynamics 177
Second Law of Thermodynamics 181
Third Law of Thermodynamics 182
Gibbs Free Energy 182
Chemical Kinetics 183
Equilibrium 188
Le Châtelier's Principle 188
Why Is Thermodynamics So Important? 189
Quiz 190

CHAPTER 12   **Organic Chemistry: All about Carbon**   **193**
What Is Organic Chemistry?   194
Carbon—More Amazing Than Ever   194
Hydrocarbons   196
Naming Organics   200
Bond Polarity   202
Common Functional Groups   203
Isomers   204
Organic Reactions   208
Quiz   211

CHAPTER 13   **Biochemistry**   **213**
What Is Biochemistry?   214
Hydrocarbons—Hydrophilic versus Hydrophobic   215
Carboxylic Acids   216
Esters   217
Amines   219
Amides   220
Phenols   222
What Are Proteins?   223
What Are Enzymes?   226
What Are Carbohydrates?   229
What Are Lipids?   231
Biological Markers   234
Nanomedicine   236
Quiz   239

CHAPTER 14   **Environmental Chemistry**   **241**
What Is Environmental Chemistry?   242
Contamination   242
What Is Acid Rain?   245
Greenhouse Effect   247
Biodegradable   253
Quiz   255

CHAPTER 15   **Nuclear Chemistry**   **257**
What Is Radioactivity?   258
Isotopes   258
Nuclear Reactions and Balance   261
What Is Radioactive Decay?   261
Radiation Detection   265
Magic Numbers   267
What Is Half-Life?   267

Radiation Exposure   271
Radiation Dosage   272
Nuclear Medicine   273
Other Radioactive Element Use   275
Radioactive Waste   275
Quiz   276

Final Exam   *279*
Answers to Quizzes and Final Exam   *301*
Appendix: SI Base Units and Conversions   *305*
Glossary   *309*
References and Internet Sites   *325*
Chemistry-at-a-Glance Study Sheets   *329*
Periodic Table   *333*
Index   *335*

Preface

Chemistry Demystified® is for anyone who is interested in chemistry and wants to learn more about this important scientific area. It can also be used by home-schooled students, tutored students, and people wanting to change careers. The material is presented in an easy-to-follow way and can be best understood when read from beginning to end. However, if you just need more information on specific topics (e.g., oxidation/reduction, enthalpy, radioisotopes) or want to brush up on organic molecules, then specific chapters can be reviewed separately.

In this second edition, I have combined some of the original chapters (e.g., solids and liquids) and added new ones (e.g., biochemistry and thermodynamics). I've updated the milestone ideas and accomplishments of chemists, biologists, physicists, and engineers to give you a sense of how the questions and ideas of people just like you advanced humankind.

Science is all about curiosity and a desire to figure out how something happens. Nobel Prize winners were once students who daydreamed about tackling problems in new ways. They knew answers had to be there and were stubborn enough to keep digging for them. Since 1901, the Nobel Prize has been awarded over 500 times for scientific excellence. The youngest person to receive the award, British physicist Lawrence Bragg, was only 25 years old when he shared the award in 1915 with his physicist father, Sir William Henry Bragg, for their work on atomic crystal structure and x-ray diffraction.

By the end of his life, Alfred Nobel had 355 patents for various inventions. After his death in 1896, Nobel's will described the establishment of a yearly international award "for those who, in the previous year, have contributed best towards the benefits for humankind" in the areas of chemistry, physics, physiology/medicine, literature, and peace. In 1968, the Nobel Prize in economics was established. Over 829 people have received the Nobel in all areas since the first prize was given out.

Nobel wanted to recognize innovative heroes and encourage others in their quest for knowledge. Perhaps by learning about past prize-winning discoveries, your own creativity will be sparked.

This book provides a general chemistry overview with sections on all the main areas you'll find in a chemistry class or an individual study of the subject. The basics are covered to familiarize you with the terms, concepts, and tools most used by scientists, physicians, and engineers. I have also listed Internet sites with intriguing up-to-date information and interactive learning devices.

Throughout the text, there are illustrations to help you visualize what is happening in chemical structure, bonding, and reactions. Quiz and final exam questions are provided. All the questions are multiple-choice and much like those used in standardized tests. Each chapter has a short "open-book" quiz. You shouldn't have any trouble with these. You can look back through the chapter to refresh your memory or check reaction details. Write down your answers and have a friend or parent check them with the answers in the back of the book. Take your time going through each chapter and don't move on until you have a good handle on the material and get most of the quiz questions right.

The final exam at the end of the book is made up of easier questions than those on the quizzes. Take the exam when you have finished all the chapter quizzes and feel comfortable with the material as a whole. A good score on the exam is at least 75% of answers correct.

With the quizzes and final exam, you may want to have your friend or parent give you your score without revealing which questions you missed. Then you might not be tempted to memorize the answers to the missed questions, but instead go back and see if you missed the point of an idea. When your quiz scores are where you'd like them to be, go back and check individual questions to confirm your strengths and any areas needing more study.

Try going through a chapter a week. An hour a day or so will allow you to take in the information slowly. Don't rush. Chemistry is not difficult, but does take some thought to decipher at times. Just plow through at a steady rate. If you want more information on buffers, spend more time in Chap. 10. If you need the latest on biological markers, allow more time in Chap. 13. After completing the course and you are a chemist-in-training, this book can serve as a ready reference guide with its glossary, Chemistry-at-a-Glance study sheets, Periodic Table, appendix, and comprehensive index.

Linda D. Williams

Acknowledgments

Illustrations in this book were generated with CorelDRAW and Microsoft PowerPoint, courtesy of the Corel and Microsoft Corporations, respectively.

National Oceanic and Atmospheric Administration (NOAA), National Aeronautics and Space Administration (NASA), Environmental Protection Agency (EPA), and United States Department of Agriculture (USDA) information was used where indicated.

A special thanks to Paul Grover Miller, Ph.D., Associate Professor, College of Medicine, University of Arkansas for Medical Sciences, for suggestions on subject rearrangement and chemistry expertise during the technical review of this book.

Many thanks to Judy Bass at McGraw-Hill for her amazing energy and support despite life's hiccups and occasional derailments. *Chemistry Demystified*, Second Edition, is a testament to her vision for the math and science books in the *Demystified* series.

About the Author

Linda D. Williams is a nonfiction writer with specialties in science, medicine, and space. Ms. Williams's work has ranged from biochemistry and microbiology to genetics and human enzyme research. With a background in microbiology and immunology, she has worked as a lead scientist and/or technical writer for Wyle Laboratories, McDonnell Douglas Space Systems, and Rice University, and served as a science speaker for the Medical Sciences Division at NASA–Johnson Space Center. Ms. Williams has more than 20 years of science research experience and has published over a dozen books, including several in the *Demystified* series (e.g., *Environmental Science Demystified*). Her work has been translated into several languages. Additionally, she has been a substitute science teacher at the elementary, intermediate, and high school levels, and founded a Science Café to share science and technology discoveries with the public.

chapter 1

Chemistry

Our ancestors didn't have readily available food, medicine, and machine-made products. Everyday life included drying fish and meats with salt, concentrating of liquids into dyes, and melting and shaping metal ores into tools. Trial-and-error testing offered clues to the makeup of the natural world. What worked was carried over by the next generation; what didn't was discarded. Many substances of the physical world were a mystery.

CHAPTER OBJECTIVES

In this chapter, you will

- Learn how the scientific method works
- Understand the International System of Units (SI)
- Find out the difference between precision and accuracy
- Understand conversion factors
- Learn why temperature is important

What Is Matter?

Chemistry is a science centered around the simple question, what is matter? Aristotle (384–322 B.C.), a student at the Greek Academy, thought matter was composed of four elements: fire, water, air, and earth. He wrote that neither form nor matter existed alone, but in hot, moist, dry, and cold combinations, which united to form the elements. Aristotle's explanation of the world was accepted for nearly 1800 years. But times changed, and so did our understanding of matter. The chapters of this book explore, step by step, the concepts about matter just as they were discovered and explained over time.

Alchemy

Aristotle's four-element theory, along with the formation of metal alloys, was the basis of early chemistry and *alchemy*.

A mixture of trickery and art, alchemy promised amazing things (e.g., lead into gold) to those who held its power. Alchemists became superstars. Those who made wild claims but couldn't deliver were permanently benched. Others made scientific progress. Crystallization and distillation of solutions began to be understood and used as standard practices. Many previously unknown elements and compounds were discovered.

Chemistry is the science of substances (i.e., matter), including structure, properties, and the reactions that change them into other substances.

What Is Modern Chemistry?

As a study of matter, chemistry is a physical science. Chemists isolate and study not just atoms and molecules, but solutions, ceramics, and metal alloys. Through experimentation, chemists study what substances do and how they react.

Chemistry is grouped into main areas or *disciplines*. (See Table 1-1.) These include

- Analytical chemistry—the use of precise instrumentation to analyze a mixture for the kinds and amounts of substances present
- Biochemistry—the study of living organisms and systems at the molecular level, including processes such as metabolism, reproduction, and digestion
- Inorganic chemistry—the study of the structure and properties of all compounds (except carbon) (e.g., salts)

TABLE 1-1 Chemistry disciplines

Discipline	Description
Agrochemistry	application of chemistry for agricultural production, food processing, and environmental remediation as a result of agriculture; also called agricultural chemistry
Analytical chemistry	material properties or developing tools to analyze materials
Astrochemistry	composition/reactions of chemical elements and molecules found in stars, in space, and interactions between matter and radiation
Biochemistry	chemical reactions taking place inside living organisms
Chemical engineering	practical application of chemistry to solve problems
Cluster chemistry	clusters of bound atoms, intermediate in size between single molecules and bulk solids
Combinatorial chemistry	computer simulation of molecules and reactions between molecules
Electrochemistry	chemical reactions in a solution at the interface between an ionic conductor and an electrical conductor
Environmental chemistry	chemistry of soil, air, and water and of human impact on natural systems
Geochemistry	chemical composition/processes associated with the Earth and other planets
Green chemistry	processes/products that eliminate or reduce the use or release of hazardous substances
Inorganic chemistry	structure and interactions between inorganic compounds (i.e., compounds not based on carbon)
Kinetics	examines the rate at which chemical reactions take place and the factors affecting the rate of chemical processes
Nanochemistry	assembly and properties of nanoscale groupings of atoms or molecules
Nuclear chemistry	chemistry associated with nuclear reactions and isotopes
Organic Chemistry	the chemistry of carbon and living things
Photochemistry	chemistry centered around interactions between light and matter
Physical chemistry	applies physics to the study of chemistry (e.g., quantum mechanics and thermodynamics)
Polymer chemistry	examines the structure and properties of macromolecules and polymers; synthesis of these molecules; also called macromolecular chemistry
Solid state chemistry	focused on the structure, properties, and chemical processes in the solid phase (e.g., synthesis and characterization of new solid state materials)
Theoretical chemistry	applies chemistry/physics calculations to describe or predict chemical events

- Organic chemistry—the study of carbon and all substances containing carbon, including most biological compounds, drugs, petroleum, and plastics
- Physical chemistry—the study of matter's physical properties and the creation of models examining why a chemical reacts in a specific way

Ancient Egyptians were the first chemists. They pioneered the art of chemistry using solutions. By 1000 B.C. ancient civilizations were using technologies that formed the basis of the various chemical disciplines. Analytical chemistry arose from extracting metal from ores, as well as chemicals from plants for medicine and perfume. Fermentation to make beer, wine, and cheese involved biochemistry. Inorganic chemistry was important in making alloys like bronze, pottery and glazes, glass, and pigments for cosmetics and paintings. The applications of organic chemistry were diverse, including the dying of cloth, tanning of leather, rendering of fat into soap, and making of organic pigments. Many of these techniques involved keen observations of the physical chemistry of compounds in order to isolate and change them for different purposes.

Basic and Applied Science

Chemistry is made up of both basic and applied science. Researchers peer into a chemical's treasure chest of secrets and try to understand why it acts the way it does. Basic science then tries to understand the rules governing the properties of matter.

However, most people know more about applied science, since it applies to everyday things. How is rust formed and how do you remove it? How do clothes get clean when washed with soap made from ashes and fat? Why does copper turn green and then black when exposed to air? How can self-assembled carbon nanotubes carry information and electricity?

The federal government supports basic and applied science through many agencies like the National Institutes of Health (NIH) and National Aeronautical and Space Administration (NASA). NASA is famous for applying basic science in new ways. NASA tests how something behaves in space with almost no gravity, like the formation of crystals or the loss of muscle tissue, and then uses that information to understand ground-based experiments.

By teaming with scientists in industry, NASA improves pharmaceuticals, optics, and bioengineering devices. Research applied in this way can more quickly travel from the laboratory to the individual. The partnering between federal institutions like NASA and industry is called *spinoffs*. A sampling of NASA's science and technology spinoffs is provided in Table 1-2.

TABLE 1-2 NASA spinoffs

Year	NASA Technology Spinoff
1976, 2005	Memory foam absorbs shock and provides extra safety in NASCAR, Formula 1, Champion Auto Racing Team (CART), and Indy Racing League racecars; motorcycle and horseback saddles; amusement park rides; military/civilian aircraft; archery targets; full-sized body casts; and human/animal prostheses
1976	Improved firefighter's breathing system lightweight firefighter's backpack system weighs only 20 lb for 30 min air supply; redesigned face mask permits better vision; personal warning device notifies firefighters when they are running out of air; reduces confusion in a fire when verbal communication is impossible
1981	Cordless tools originally came from development of a portable self-contained drill able to extract core samples up to 10 ft below the surface of the moon; now cordless technology used in everything from weed eaters to orthopedic instruments
1984, 1996	Scratch-resistant lenses with diamond-hard coatings are scratch-resistant and shed water more easily, reducing spotting
1994	LORAD stereo guide breast biopsy system uses a digital camera system to replace surgical biopsy and is performed under local anesthesia with a needle and saves women time, pain, scarring, radiation exposure, and money
1995, 2008	Microbial check valve; iodine-dispensing to purify drinking water
1995, 2009	Ethylene reduction device keeps produce and plants from maturing too fast by converting ethylene (C_2H_4) into water (H_2O) and carbon dioxide (CO_2)
2003	Humanitarian demining device is a flare-like device that uses rocket fuel to create a high-temperature flame, which burns out the explosive fill in landmines before they can explode
2006	Eagle Eyes lens protects human vision from the detrimental effects of radiation from the light spectrum
2002	Automatic implantable cardiovertor defibrillator monitors the heart continuously, recognizes the onset of ventricular fibrillation, and delivers a corrective electrical shock
2002	Virtual Window provides real-time 3D images without glasses, head trackers, helmets, or other viewing aids
1977, 2003	Cochlear implant selects speech signal information and then produces a pattern of electrical pulses in a patient's ear. A microphone picks up sounds and transmits them to a speech processor that converts them into digital signals
2006	Petroleum remediation product has thousands of microcapsules (i.e., tiny hollow balls of beeswax); water can't enter the microcapsule's cell, but oil is absorbed into the beeswax spheres as they float on the water's surface
2005, 2008	Light-emitting diodes (LED) for medical applications such as cancer treatment
2009	Givens Buoy Life Raft has a heavy, water-filled ballast to keep the center of gravity constant making the raft nearly impossible to capsize even in 100 mi/h winds
2009	Bioreactor to multiply stem cells from a patient's blood to treat disease
2009	Carbon nanotube biosensor alerts inspectors to minute amounts of dangerous organic contaminants and pathogens

NASA spinoffs include computer technology, consumer products for recreation and the home, environmental and resource management, industry and manufacturing, public safety, and transportation.

The keys to the scientific method are curiosity and determination, observation and analysis, measurement and conclusion. As humans, we are curious by nature. In the following chapters, you'll see how scientists satisfy this curiosity.

Scientific Method

The early development of the scientific method arose from Aristotle's laws of logic. He saw the importance of observation and then classified what was observed in order to better understand nature.

In the Middle Ages, Ibn al-Haytham, a Persian mathematician and student of Greek philosophy, developed the scientific method further. His study of Aristotle's works made him realize that physical science and mathematics were important keys to unlocking the universe's mysteries. During his life, he developed different experiments to check his physical observations and made valuable discoveries in the study of vision. Al-Haytham's seven-volume *Book of Optics*, written between A.D. 1011 and 1021, correctly described the transmission, reflection, and refraction of light. His work demonstrated the early power of the scientific method.

In modern times, Galileo (1564–1642) is commonly credited with being the father of the scientific method although many scientists have added to his understanding over the centuries. By the twentieth century, the scientific method was arranged into four steps:

- Observation
- Hypothesis
- Prediction
- Experimentation

Ever since fire was first discovered, people noticed how it changed its environment; nearby grass was burned and trees charred. Eventually, by following the scientific method, scientists made great discoveries, such as what happens when something burns. They realized collecting as much information as possible before any conclusions was critical to gaining understanding.

In the eighteenth century, Antoine Lavoisier, a French scientist, found that when silvery mercury was burned in the air, it turned into a red-orange substance with a greater mass than that of an original mercury sample. He also made observations about the gases in the air. For these discoveries, Lavoisier is often called the father of chemistry.

Hypothesis

When all facts are known, the next step in the scientific method is to develop a hypothesis. A hypothesis is a statement explaining an observation.

Lavoisier created a *hypothesis* based on his observations with fire and air. He proposed a hypothesis to explain *combustion* or how things burn. His idea was that some part of air combined with a burning sample and transformed it. Lavoisier called this mystery part oxygen.

Hypothesis is a statement that describes or explains an observation.

A hypothesis is important not just to explain what is seen, but also to predict what might happen. If something in the air combines with a sample, then the new substance (i.e., formed after burning) should have added mass from air's contribution to the combustion. It also should be possible to reverse the process.

? Still Struggling

Lavoisier knew how important it was to carry out accurate experiments. He showed after burning mercury that a new, heavier substance, mercury oxide, was formed. He also showed the reverse reaction (i.e., mercury's original mass could be regained). In other words, oxygen could be reclaimed from the mercury oxide.

An **experiment** is a controlled testing of a substance or system's properties by carefully recorded measurements.

Now to learn if a hypothesis is true, it must be tested with *experiments*. Lavoisier's additional experiments showed that air is made up of several other gases (e.g., nitrogen), but unlike oxygen, they didn't combine with mercury. His hypothesis on combustion became a *theory*, which is a hypothesis thoroughly proven by experimentation.

> A **theory** is the result of thorough testing and the confirmation of a hypothesis.

Following later experimentation by other scientists in many different disciplines such as astronomy, electricity, mathematics, biology, chemistry, and medicine, data was recorded which supported how almost everything could be studied and predicted through a series of observations and calculations. When scientists around the world got the same results repeatedly, a particular hypothesis or theory became a *law*.

> A **scientific law** is a hypothesis or theory that is tested time after time with the same resulting data and thought to be without exception.

Measurements

Observation and measurement, as in all of science, are the keys to chemistry. In research, as in other parts of life, we are constantly measuring. The baseball cleared the outfield fence by a foot. The soccer ball missed the flowerpot by 3 inches (in). The Austrian driver cruised at 160 kilometers per hour (km/h). The Kentucky Derby favorite pulled ahead by a length. The Olympic skier slid into first place by two one-hundredths of a second. The soldier's letter home weighed 1 ounce (oz).

Research is all about measuring. To repeat an experiment or follow someone else's method, the *same* units must be used. It doesn't work to have a researcher in New York measuring in cups while another in Germany measures in milliliters. To repeat an experiment and learn from it, scientists worldwide needed a common system.

In 1670, a French scientist named Gabriel Mouton suggested a *decimal system* of measurement. Units were based on groups of 10. It took a while for people to try it for themselves, but in 1799 the French Academy of Sciences established a decimal-based system of measurement. They called it the *metric system*, from the Greek *metron*, which means a measure. On January 1, 1840,

TABLE 1-3 SI base units are used in chemistry		
Measurement	**Unit**	**Symbol**
Mass (not weight)	kilogram	kg
Length	meter	m
Temperature	kelvin	K
Time	second	s
Pure substance amount	mole	mol
Electric current	ampere	A
Light brightness (wavelength)	candela	cd

the French Legislature passed a law requiring the metric system be used in all trade.

International System of Units (SI)

In 1960, the General Conference on Weights and Measures adopted the *International System of Units* (*SI*), from Le Système International d'Unités. The International Bureau of Weights and Standards in Sèvres, France, has the official platinum standard measures by which all other standards are compared. The SI system has *seven base units* from which other units are calculated. Table 1-3 lists SI units used in chemistry.

When Great Britain formally adopted the metric system in 1965, the United States became the only major nation that didn't require metric, though people had been using it since the mid-1800s.

Scientific Notation

Scientific notation is a simple way to write and keep track of large and small numbers without tons of zeros. It offers a shortcut way to do calculations and record results.

The SI measuring system is based on a decimal system. With calculations written in groups of 10, results can be easily recorded as exponents of 10. Written superscripts indicate exponential values. For example, 1,000,000 can be written as 1×10^6 or 0.001 can be written as 1×10^{-3}. (Note: 1×10^0 is just 1.) Some of the terms used in exponential notation are listed in Table 1-4.

The ease of this method is shown below.

TABLE 1-4 Exponential notation		
Prefix	**Symbol**	**Value**
tera	T	10^{12}
giga	G	10^{9}
mega	M	10^{6}
kilo	k	10^{3}
deca	da	10^{1}
deci	d	10^{-1}
centi	c	10^{-2}
milli	m	10^{-3}
micro	μ*	10^{-6}
nano	n	10^{-9}
pico	p	10^{-12}
*Greek letter mu		

EXAMPLE 1-1

$$100 = (10)(10) = 10^2 = \text{one hundred}$$

$$1000 = (10)(10)(10) = 10^3 = \text{one thousand}$$

$$10,000 = (10)(10)(10)(10) = 10^4 = \text{ten thousand}$$

$$100,000 = (10)(10)(10)(10)(10) = 10^5 = \text{one hundred thousand}$$

$$1,000,000 = (10)(10)(10)(10)(10)(10) = 10^6 = \text{one million}$$

$$1,000,000,000 = (10)(10)(10)(10)(10)(10)(10)(10)(10) = 10^9 = \text{one billion}$$

$$1,000,000,000,000 = (10)(10)(10)(10)(10)(10)(10)(10)(10)(10)(10)(10)$$
$$= 10^{12} = \text{one trillion}$$

$$1/10 = 10^{-1} = \text{one tenth}$$

$$1/100 = 1/(10)(10) = 10^{-2} = \text{one hundredth}$$

$$1/1000 = 1/(10)(10)(10) = 10^{-3} = \text{one thousandth}$$

$$1/10,000 = 1/(10)(10)\,(10)(10) = 10^{-4} = \text{one ten thousandth}$$

$$1/1,000,000 = 1/(10)(10)(10)(10)(10)(10) = 10^{-6} = \text{one millionth}$$

$$1/1,000,000,000 = 1/(10)(10)(10)(10)(10)(10)(10)(10)(10) = 10^{-9}$$
$$= \text{one billionth}$$

$$1/1,000,000,000,000 = 1/(10)(10)(10)(10)(10)(10)(10)(10)(10)(10)(10)(10)$$
$$= 10^{-12} = \text{one trillionth}$$

TABLE 1-5	In science, metric units are used rather than English units	
Sample		**Measurement (meters)**
Diameter of uranium nucleus		10^{-13}
Water molecule		10^{-10}
Protozoa		10^{-5}
Earthworm		10^{-2}
Human		2
Mount Everest		10^3
Diameter of the Earth		10^7
Distance from Pluto to the Sun		10^{13}

Since the *English System* was used in the United States for many years with units of inches, feet, yards, miles, cups, quarts, gallons, etc. many people were not comfortable with the metric system until recently. Most students wonder why anyone ever preferred the older system when they discover how easy it is to multiply metric units. Table 1-5 lists some everyday metric measurements.

Significant Figures

Measurements are never exact, but scientists try to record an answer with the least amount of uncertainty. Scientific notation, then, was set up to standardize measurements. The idea of *significant figures* was created to write numbers either in whole units or to the highest level of confidence.

Significant figures are the number of digits in which there is high confidence in the value.

A *counted* significant figure is something that cannot be divided into subparts. These are written in whole numbers such as 9 horses, 2 snowboards, or 7 keys. *Defined* significant digits are exact numbers, but not always whole numbers [e.g., 2.54 centimeters (cm) equals 1 in].

Helpful Rules of Significant Figures:
1. All nonzero digits (1, 2, 3, etc.) = significant.
2. Zeroes between nonzero digits = significant.
3. Zeroes on the left of the first nonzero digit = not significant.

4. Zeroes on the right of the last nonzero digit = significant with a decimal point. (No decimal point, not significant.) (0.705200 has six significant digits, while 705,200 has only four.)

5. When using scientific notation, the number of digits shown is the number of significant digits (7.052×10^5). This example contains four significant digits.

EXAMPLE 1-2

How many significant figures are in the following?

(a) 9.107 (four, zero in the middle is significant)
(b) 401 grams (g) (three, zero in the middle is significant)
(c) 0.006 (one, leading zeros are never significant)
(d) 800 km (three, zeros are significant in measurements unless otherwise indicated)

SOLUTION

Remember, when finding the number of significant figures, check for zeros acting as placeholders. Leading zeros to the left-hand side of a number are never significant. You start at the left and count to the right of the decimal point. The measurement 0.096 m has two significant figures. The measurement 13.42 cm has four significant figures. The mass 0.0027 g has two significant figures. (Note: Remember to leave off the leading zeros.)

Sandwiched (in the middle) zeros are always significant. The number 26,304 has five significant figures. The measurement 0.000001002 m has four significant figures.

Scientific notation gets rid of guessing and helps to keep track of zeros in very large and very small numbers. If the diameter of the Earth is 10,000,000 meters (m), it is more practical to write 1×10^7 m. Or, if the length of a virus is 0.00000004 m, it is easier to write 4×10^{-7} m.

When multiplying or dividing numbers, the significant digits of the number with the least number of significant digits shows the number of significant digits in the answer.

EXAMPLE 1-3

40 lb potatoes × $.45 per lb = $18.00 or $18 since the first number is only measured to ones place.

EXAMPLE 1-4
0.5 oz of perfume × $25.00 per oz = $12.50 for 0.5 oz of perfume.

EXAMPLE 1-5
6.23 ft of wood × $2.00 per linear ft = $12.46 per linear ft.

Precision versus Accuracy

Imagine you are trying to prove a theory based on a chemical property like boiling point. You have to be certain of the boiling point. If the boiling point temperature was loosely recorded by others, you'll have trouble repeating the experiment. Because theories become laws by repeated experimentation, it is important to record measurements precisely.

Scientific knowledge moves forward by building on results and experiments done by earlier scientists. *Precision* is directly related to measurement's reproducibility. If previous measurements were sloppy, a researcher wouldn't know if the current results were new and exciting or just wrong.

> **Precision** describes how a measurement can be reproduced time after time.

Closely related to the topic of precision is that of *accuracy*. Some people use the two interchangeably, but there is a big difference between them.

> **Accuracy** describes the closeness of a measurement to its true value.

? Still Struggling

In baseball, when player A throws balls at a target's center, it represents high precision and accuracy. Player B's aim, with balls high and low missing the target, represents low precision and low accuracy. Player C's hits, clumped together at the bottom left side of the target, define high precision (since they all landed in the same place), but low accuracy (since the object is to hit the target's center). Player C, then, has to work on hitting the target's center, if he wants to win games and improve accuracy.

Rounding

Rounding is the way to drop (or leave out) nonsignificant numbers in a calculation and adjust the last number up or down.

There are three basic rules to remember when rounding numbers:

1. If a digit is ≥ 5 followed by nonzeros, add 1 to the last digit. (Note: 3.2151 is rounded to 3.22.)

2. If a digit is < 5, then the digits are dropped. (Note: 7.12132 would be rounded to 7.12.)

3. If the number is 5 (or 5 and a bunch of zeros), round to the least certain number of digits. (Note: 4.823, 4.82309, and 4.8150 all round off to 4.82.)

Rounding reduces accuracy, but increases precision. The numbers get closer, but are not necessarily on target.

EXAMPLE 1-6

Try rounding the numbers below for practice.

(a) 2.2751 to three significant digits
(b) 4.114 to three significant digits
(c) 3.177 to two significant digits
(d) 5.99 to one significant digit
(e) 2.213 to two significant digits
(f) 0.0639 to two significant digits

Did you get (a) 2.28, (b) 4.11, (c) 3.2, (d) 6, (e) 2.2, and (f) 0.064?

SOLUTION

When multiplying or dividing measurements, the significant digits of the measurement with the *least* number of significant digits provides the number of significant digits in the answer.

EXAMPLE 1-7

Do you see how significant digits are rounded up or down?

(1) 1.8 lb of oranges × $3.99 per lb = 7.182 = $7.18 = $7.2 (Note: 1.8 lb of oranges has one significant digit.)

(2) 15.2 oz of olive oil × $1.35 per oz = $20.52

(3) 25 linear ft of rope × $3.60 per linear ft = $90.00

☑ **SOLUTION**

Measurements are calculated to a high precision. Calculators can provide 8 to 10 numbers for a calculation, but most measurements require far less accuracy.

Rounding makes numbers easier to work with and remember.

Think of how out-of-town friends would react if you said, "drive 3.4793561 miles (mi) west on Main Street; turn right and go 14.1379257 mi straight until Union Street; turn left and travel 1.24900023 mi around the curve until you reach the red brick house on the right."

They might never arrive! But, by rounding to 3.5 mi, 14 mi, 1.2 mi, and watching the car's odometer (the instrument that measures distance), your friends would arrive with a lot less confusion.

Conversion Factors

Dimensional Analysis (Factor-Label Method)

Conversion factors use a relationship between units or quantities given in fractional form. The *factor-label method* (also known as *dimensional analysis*) changes one unit into another by using conversion factors.

Conversion factors are helpful when you want to compare two measurements that aren't in the same units.

If given a measurement in meters and the map reads only in kilometers, you have a problem. You could guess or use the conversion factor of 1 km/10^3 m. Look at the following conversion:

$$0.392 \text{ m} \times 1 \text{ km}/10^3 \text{ m} = 0.392 \times 10^{-3} \text{ km}$$

$$= 3.92 \times 10^{-4} \text{ km}$$

If you have centimeters and need to know the answer in inches, then use a conversion factor of 1 in/2.54 cm.

914 cm × 1 in/2.54 cm = 360 in (since 914 has three significant digits)

Converting measurements can also be a two-step process.

$$mg \rightarrow g \rightarrow kg$$
$$L \rightarrow qt \rightarrow gal$$
$$mi/h \rightarrow m/min$$

Look at the following two-step conversions.

EXAMPLE 1-8

$$2461 \text{ mg} \rightarrow \text{kg}$$
$$\text{mg} \rightarrow \text{g} \rightarrow \text{kg}$$
$$1 \text{ mg} = 10^{-3} \text{ g}; 1 \text{ kg} = 10^{3} \text{ g } (\textit{conversion factors})$$
$$2461 \text{ mg} \times 10^{-3} \text{ g/mg} \times 1 \text{ kg/} 10^{3} \text{ g} = 2461 \times 10^{-6} \text{ kg} = 2.461 \times 10^{-3} \text{ kg}$$

EXAMPLE 1-9

$$8.74 \text{ L} \rightarrow \text{gal}$$
$$\text{L} \rightarrow \text{qt} \rightarrow \text{gal}$$
$$1.06 \text{ qt/L and 1 gal/4 qt } (\textit{conversion factors})$$
$$8.47 \text{ L} \times 1.06 \text{ qt/L} \times 1 \text{ gal/4 qt} = 2.24 \text{ gal}$$

EXAMPLE 1-10

$$70 \text{ mi/h} \rightarrow \text{m/min}$$
$$\text{mi/h} \rightarrow \text{km/h} \rightarrow \text{m/h} \rightarrow \text{m/min}$$
$$1.61 \text{ km/mi}; 10^{3} \text{ m/km}; 1 \text{ h/60 min } (\textit{conversion factors})$$
$$70 \text{ mi/h} \times 1.61 \text{ km/mi} \times 10^{3} \text{ m/km} \times 1 \text{ h/60 min} = 1878.33 \text{ m/min} =$$
$$1.9 \times 10^{3} \text{ m/min}$$

SI derived units are obtained by combining SI base units.

Celsius (°C) to Fahrenheit (°F) to Kelvin (K)

The measure of the heat intensity of a substance is said to be its *temperature*. A thermometer measures temperature. Temperature is measured in three different units: *Fahrenheit* (°F) in the United States, *Celsius* (°C) in science and elsewhere, and *kelvin* (K) to measure absolute temperature. (Note: Kelvin measurements don't use a degree symbol.) For example, water boils at 212°F, 100°C, and 373 K. Likewise, water freezes at 32°F, 0°C, and 273 K. Figure 1-1 shows how the temperature systems compare.

The conversion factor for Celsius to Fahrenheit is

$$t \, (°F) = [t \, (°C) \times 1.8°F/1°C] + 32 = [t \, (°C) \times 1.8] + 32$$

The conversion factor for Fahrenheit to Celsius is

(*Hint*: Subtract 32 so that both numbers start at the same temperature.)

$$t \, (°C) = [t \, (°F) - 32°F] \times 1°C/1.8°F = [t \, (°F) - 32]/1.8$$

or a simpler way to state it is

$$°C = 5/9 \, (°F - 32)$$

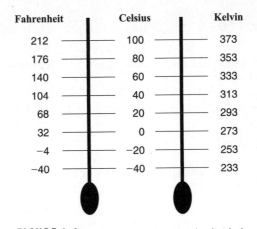

FIGURE 1-1 • A kerotakis device used a divided chamber with a furnace to provide heat.

EXAMPLE **1-11**

A summer day in Hawaii might be 21°C. What is that in Fahrenheit?

$$21°C = 5/9 (°F - 32)$$
$$21 + 32 = 5/9°F$$
$$53 \times 9 = 5°F$$
$$477/5 = 95.4°F$$

To find *absolute zero* (the lowest temperature possible), the *kelvin* scale is used, where the lowest temperature is zero. A kelvin is an SI temperature unit. The heat energy is zero.

To see how temperature conversion works, let's convert normal body temperature 98.6°F to Celsius.

EXAMPLE **1-12**

$$°C = 5/9 (°F - 32)$$
$$°C = 5/9 (98.6°F - 32)$$
$$= 5/9 (66.6) = 37.0°C$$

°C can be converted to K, by adding 273 to the Celsius temperature.

EXAMPLE **1-13**

$$K = °C + 273$$
$$K = 37°C + 273 = 310 \text{ K}$$

QUIZ

1. Who discovered that air was composed of several different components including oxygen?
   A. Aristotle
   B. Lavoisier
   C. Dobbins
   D. Copernicus

2. When alchemists changed a common metal's color, they claimed it had been transmuted into
   A. wood.
   B. iron.
   C. gold.
   D. mercury.

3. A hypothesis is a
   A. a highly accurate hypodermic syringe.
   B. way to describe temperature transfer between reactants.
   C. way to explain poor data collection.
   D. statement or idea that attempts to explain observable results.

4. A chemical experiment is
   A. an uncontrolled measurement of a mixtures' variables.
   B. a one-time reporting of a few observable characteristics.
   C. a controlled testing of a substance's properties through carefully recorded measurements.
   D. not important if a hypothesis is well thought out.

5. Because of his early work, who is often called the founder of the scientific method?
   A. Galileo Galilei
   B. Michael Caputo
   C. Antoine Lavoisier
   D. Charles Dalton

6. A theory
   A. is little better than an imaginative guess.
   B. may differ from one experiment to another.
   C. is seldom tested once it has been established.
   D. predicts the outcome of new testing based on past experimental data.

7. The International System of Units (SI) uses how many base units?
   A. 2
   B. 5
   C. 7
   D. 10

8. **Scientific notation is an important**
    A. way to write very large and very small numbers.
    B. shorthand method of number accounting.
    C. method where numbers are written in powers of 10.
    D. all of the above.

9. **Accuracy is**
    A. how close a single measurement is to its true value.
    B. the closeness of two sets of measured values.
    C. more precise than four significant digits.
    D. only applicable to experimental measurements.

10. **Dimensional analysis**
    A. only works with quantities written as whole numbers.
    B. changes one unit to another by using conversion factors.
    C. requires a direct connection between weight and volume.
    D. always compares measurements in the same units.

chapter *2*

Atomic Structure and Theory

This chapter deals with atoms (the basic units of matter), subparts (protons, electrons, and neutrons) of atoms, and atomic theory. We will also discuss short-cut structural and molecular formulas for chemical compounds.

CHAPTER OBJECTIVES

In this chapter, you will

- Learn the parts of an atom
- Understand atomic theory
- Find out which atomic particles have electrical charge
- Discover the difference between molecular and structural formulas

What Are Atoms?

Around 495 B.C. a Greek philosopher named Democritus wondered if substances could be divided into smaller and smaller parts indefinitely. He thought that eventually particles would reach a point where they could no longer be divided. He called these smallest particles *atoms* (from the Greek word, *atomos*, which means, "not divided"). This idea was further developed and discussed by the great philosophers, Aristotle and Plato, who thought matter was similarly indivisible as well as fluid and endless.

Beginnings of Atomic Theory

In 100 B.C., another forward thinker named Lucretius wrote a long, descriptive poem called *De Rerum Natura* (The Nature of Things) praising early ideas that leaned toward an *atomic theory* of matter. The telling and retelling of this poem helped people understand the basic nature of matter, because few people could read and most learned through story telling. Later, in 1452, the Gutenberg printing press, using olive oil for ink and a screw-type wine press, was invented and further spread the knowledge of the atomic theory. In fact, *De Rerum Natura*, one of the earliest texts set in print, helped the atomic theory survive over the centuries. Along with religious texts and Bibles of the time, the poem was one of the few things on hand to be read. By dispersing the atomic theory, future scientists were able to think about matter in particle form and study how it could be divided, rearranged, and combined.

Parts of Atomic Structure

Although atoms are basic units of matter, modern scientists have discovered particles smaller than an atom. These *subatomic* particles consist of nucleons and electrons. There are two types of nucleons: *protons* with a positive charge and *neutrons* with zero charge (i.e., neutral). Protons and neutrons interact to form a nucleus. *Electrons*, orbiting the nucleus like untamed satellites, are attracted by the forces of electromagnetism. Figure 2-1 shows how the core of an atom might look if a model were made of subatomic particles. This modern model was first proposed by Niels Bohr in 1913.

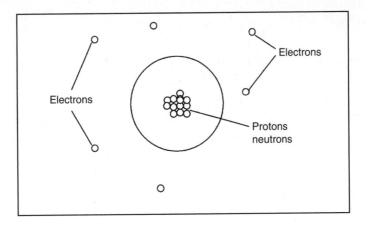

FIGURE 2-1 • The nucleus is thought to have orbiting electrons.

Electrons

In 1897, J. J. Thomson, a physician from New Zealand, discovered negatively charged particles by removing all the air from a glass tube connected to two electrodes. One negatively (−) charged electrode called the *cathode* was attached to one end of the tube, while a positively (+) charged electrode called the *anode* was attached to the other end of the tube. The resulting *cathode ray tube* used an electrical current to excite the atoms of different gases within the tube. Electricity was beamed down the length of a tube to the other electrode. By using this piece of equipment a century ago, scientists began separating the individual particles of atoms.

Through Thomson's experiments with different-colored gases, he found that *electrons* had a negative charge (−) and seemed to be common to all elements. However, Thomson's results also showed that an atom's overall charge was neutral. He thought an equal balance of positive and negative charges must cancel each other out to make the overall atom neutral.

Electrons are small negatively charged subatomic particles that orbit around an atom's positively charged nucleus.

Thomson came up with the "plum-pudding" model of subparticle arrangement made up of a mass of positively (+) charged particles, the "pudding," and specks of negatively (−) charged particles floating around in it like "raisins." He probably ate dessert right before or after working in his lab, so the idea came to

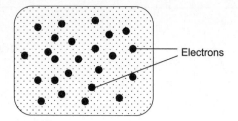

FIGURE 2-2 · The plum-pudding model of electrons and protons was not compact.

him fairly easily. The plum-pudding model of electrons and protons is shown in Fig. 2-2. In 1906, Thomson was awarded the Nobel Prize in Physics for his research and electrical work with gases.

It wasn't until a student of Thomson's, Ernest Rutherford, started working to support his teacher's ideas that the data for a plum-pudding model just didn't hold up. The floating negatively charged "raisins" acted differently in a current, for different elements, than what Thomson expected. This seemed to suggest they had different energy levels. (Maybe that is where we get the "the proof is in the pudding" expression.) In fact, other experimental data didn't quite make sense until scientists discovered that the atom was not just a solid core, but made up of smaller subparticles.

Nuclear Model

Rutherford developed the modern atomic concept and received the Nobel Prize for Chemistry in 1908. Moreover, he was knighted in 1914 for his work. (Whoever said chemistry was not a glory science?) Through his experiments with radioactive uranium in 1911, Rutherford described a *nuclear* model. By bombarding particles through thin gold foil, he predicted that atoms had positive cores much smaller than the rest of an atom.

Instead of thinking that atoms were the same all the way through ("plum-pudding" model) as Thomson suggested, Rutherford's experiments pointed toward something more like a fruit with a small, dense center. Along with those of his student, Hans Geiger, Rutherford's experiments showed that over 99% of the bombarded particles passed easily through the gold, but a few (1 out of 8000) ricocheted at wild angles and backwards. Rutherford thought this scattering happened when (+) nuclei of the test particles collided and were repelled by heavy positively charged gold nuclei. The rarity of the collision meant that there was a "dense pit" or nucleus at the center of a mostly empty atom. Figure 2-3 shows how Rutherford imagined a dense pit nuclear model.

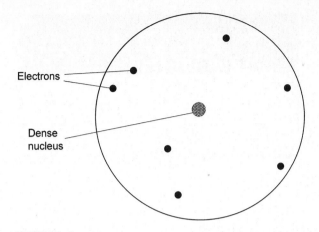

FIGURE 2-3 · Rutherford's model of the nucleus had a tight central core.

Later studies proved Rutherford's model was correct. The positively charged nucleus of an atom contains most of the mass and is very dense, but takes up only a tiny part of its total space. To get an idea of size, if an atomic nucleus were the size of a ping-pong ball, then the rest of the atom with its encircling negatively charged electrons would measure over 2 1/2 miles across. More precisely, nuclei are roughly 10^{-12} meters in diameter. The total diameter of an atom is around of 10^{-8} meters or roughly a 10,000 times larger.

Proton

A *proton* is a subatomic particle, within the nucleus of an atom. It has a positive charge and its mass is roughly 1800 times greater than that of an electron.

> **Protons** are positively charged particles in the nucleus of an atom.

The number of protons in an atom's nucleus is an element's *atomic number* (Z) as shown in the Periodic Table. To obtain an element's atomic number, just count the number of protons in the nucleus.

EXAMPLE 2-1

What is the atomic number of (a) sodium, (b) gold, (c) titanium, (d) rutherfordium, and (e) krypton?

SOLUTION 2-1

Did you get (a) 11, (b) 79, (c) 22, (d) 104, and (e) 36?

TABLE 2-1	Common characteristics of electrons, protons, and neutrons indicate their special nature	
Name	**Symbol**	**Mass (g)**
Electron	e⁻	9.110×10^{-28}
Proton	p⁺	1.675×10^{-24}
Neutron	n	1.675×10^{-24}

Neutron

Neutrons are another subatomic particle (i.e., nucleon) within an atom's nucleus. They were discovered when chemists made calculations based on atomic weights of atoms and the numbers didn't add up. The scientists knew that there must be something missing in the equation. To make up the difference in mass, it was hypothesized that nuclear particles without a charge were located inside the crowded nucleus along with positively charged protons.

Neutrons are subatomic particles with a similar mass to their partner proton in the nucleus, but with no electrical (+ or −) charge.

Table 2-1 shows common characteristics of electrons, protons, and neutrons.

Molecules

Despite simple names, many common forms of matter like wood, rock, or soap are a complex combination of atoms in a specific geometrical alignment. The force binding two or more atoms together is known as a *chemical bond*.

A *molecule* forms when two or more atoms link together by chemical bonding. In a *covalent bond*, electrons are shared equally and in an *ionic bond*, electrons are transferred and the resulting opposite charges attract. The resulting mixture of different atoms is a called a *molecular compound*. Water is a familiar molecule composed of two atoms of hydrogen and one atom of oxygen held together by covalent bonds.

> A **molecule** is composed of atoms chemically bonded by attractive forces.

In some cases, molecules are made up of particles having the same atomic number and are pure elements. Several pure elements occur naturally as two atom or *diatomic* molecules. Of these, oxygen, nitrogen, hydrogen, fluorine, chlorine, bromine, and iodine occur in pairs at room temperature. These are grouped in sections IA and VIIA of the Periodic Table (Chap. 3). Other molecules such as elemental phosphorus (P) are composed of four atoms, while sulfur (S) has eight atoms. Knowing the normal state of an element becomes important in predicting the outcome of chemical reactions.

Molecular Formulas

A compound's name doesn't offer much information about its chemical and physical properties. So, scientists use molecular formulas to discuss particular compounds.

A *molecular formula* indicates the type and number of atoms that make up a compound. This chemical shorthand helps identify a particular compound. The formula for water is H_2O. The number of atoms for each element is written as a subscript in the formula, such as the "2" in H_2O. When no subscript is written, it is understood that only one atom of the element is present (e.g., oxygen in water). With this system, scientists can tell what atoms are found in water and can predict its properties, such as hydrogen bonding ability.

Many commonly used compounds are written as molecular formulas. The molecular formula for hydrogen peroxide is similar to water, except that it has two hydrogen atoms and two oxygen atoms (H_2O_2). Ethanol (i.e., alcohol) is written as (C_2H_6O). Saltpeter (e.g., fireworks and fertilizer) is KNO_3 and fructose (i.e., fruit sugar) is $C_6H_{12}O_6$. Later, when we take a closer look at how atoms and elements combine, the importance of subscripts will become obvious. Table 2-2, lists some common molecular formulas. The following example shows a few molecular ratios for different compounds.

EXAMPLE 2-2

Some simple chemical formulas:

- Sodium chloride (NaCl) = 1 atom of sodium and 1 atom of chlorine are bonded.

TABLE 2-2 Molecular formulas show the number of atoms present in a molecule

Name	Chemical formula
Ethane	C_2H_6
Carbon tetrachloride	CCl_4
Oxalic acid	$H_2C_2O_4$
Cupric nitrate	$Cu(NO_3)_2$
Diphosphorus trioxide	P_2O_3
Ammonium nitrate	NH_4NO_3
Urea	NH_2CONH_2
Sulfuric acid	H_2SO_4
Calcium hydroxide	$Ca(OH)_2$
Sodium stearate	$C_{18}H_{35}O_2Na$
Benzene	C_6H_6

- **Ammonia (NH_3) = 1 atom of nitrogen and 3 atoms of hydrogen are bonded.**
- **Sucrose ($C_6H_{12}O_6$) = 6 atoms of carbon, 12 atoms of hydrogen, and 6 atoms of oxygen are chemically bonded.**

A **molecular formula** gives the exact number of an element's different atoms in a molecule.

Structural Formulas

Scientists also use *structural formulas* to draw compounds. While a molecular formula shows a compound's make-up, it gives little information about how the atoms are connected in space. Based on accepted standards, a structural formula illustrates chemical bonds between atoms.

Bonding happens according to various elemental properties and the location of electrons around a nucleus. The ways in which atoms are connected often determine a compound's chemical and physical properties. Scientists study these relationships to improve understanding and to predict how compounds will act in different chemical reactions.

Structural formulas show how specific atoms are arranged and bonded in a compound.

? Still Struggling

Think of it like a football game. The plays are set up with different players placed in certain positions. Each play is designed to serve a particular purpose. If the players line up one way, the quarterback may throw the ball. Set up another way and the end player runs the ball over and across. If the players on the other side don't react to a certain configuration in a certain way, the quarterback may have to run the ball. Placement and function of individual players is everything in football.

The same is true of chemistry. Atoms' arrangement in a molecule can make a big difference in the characteristics and reactivity of compounds. Figure 2-4 shows a variety of structural formulas with individual elements indicated.

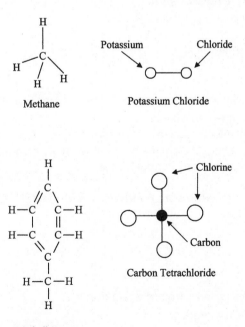

FIGURE 2-4 · Structural formulas of different compounds make it easier to see how atoms are bonded.

QUIZ

1. The molecular formula for hydrogen peroxide is
    A. H_2O
    B. H_2O_2
    C. H-O-H
    D. HO_2

2. A proton has roughly
    A. the same mass as an electron.
    B. double the mass of an electron.
    C. roughly 1800 times greater mass than an electron.
    D. 2000 times smaller mass than an electron.

3. What is composed of atoms chemically bonded by attractive forces?
    A. Electron
    B. Proton
    C. Neutron
    D. Molecule

4. Neutrons share the nucleus with
    A. electrons.
    B. molecules.
    C. carbon nanotubes.
    D. protons.

5. Thomson's experiments with several different-colored gases showed
    A. sodium burns green in a flame.
    B. magnetic spin affects an element's reactivity.
    C. electrons had a negative charge and seemed to be common to all elements.
    D. the limitation of the early Periodic Table.

6. How many hydrogen atoms are bonded in a sucrose ($C_6H_{12}O_6$) molecule?
    A. 0
    B. 6
    C. 12
    D. 24

7. The neutron gets its name from the fact that it
    A. has no electrical charge.
    B. has a slight positive charge.
    C. is more reactive than an electron.
    D. is smaller than a neutrino.

8. **A molecular formula**
   - A. provides the spatial orientation of atoms in a molecule.
   - B. gives the number of each elemental atom in a molecule.
   - C. gives the total number of quarks in a compound.
   - D. is not used in environmental chemistry.

9. **What is the atomic number of tin (Sn)?**
   - A. 18
   - B. 27
   - C. 50
   - D. 62

10. **A structural formula provides information**
    - A. only for organic compounds.
    - B. to calculate a solution's boiling point.
    - C. on an element's color at room temperature.
    - D. on how an element is spatially bonded in a molecule.

chapter 3

Elements and the Periodic Table

Matter is the "stuff" that everything is made of. In this chapter, we will discuss the three main forms of matter and their properties. We will look at the characteristics of the elements and learn how to use the Periodic Table.

CHAPTER OBJECTIVES

In this chapter, you will

- Learn about the different forms and properties of matter
- Understand how chemical naming works
- Learn how elements fit into the Periodic Table
- Find out the difference between atomic number and atomic weight
- See what factors place elements into different classes, periods, and groups

What Is Matter?

Matter makes up everything you see, touch, smell, taste, or can measure (and some you can't). Bubble gum to down pillows, bamboo to rocket fuel are all different forms of matter.

Three Main Forms

Matter generally takes one of three forms, solid, liquid, or gas, but matter may also change from one form to another and back again. For example, water is liquid at room temperature, solid when frozen, and a gas when boiling.

Properties

Matter has specific physical and chemical properties. Some types of matter are aligned in crystalline structures, while others exist mainly in the gaseous state. Some matter is extremely reactive and volatile, while other types are quite stable and may not react at all unless under extreme temperature or pressure. We will study these characteristics of matter in more depth in Chaps. 4 and 5.

Elements

For thousands of years, people have known about the basic characteristics of the elements. They knew rocks were hard, rivers were liquid, and fog was a mist. They knew some materials could be packed down, heated, frozen, and altered in different ways, while others could not. What they didn't understand was how matter seemed to turn from one form into another. For example, why iron changed into rust was a mystery.

Although, initially more myth than science, experience-based explanations eventually provided a deeper understanding. As we learned in Chap. 1, Antoine Lavoisier insisted on precise measurements to better compare results and explain the properties of matter. He defined an *element* as a substance that could not be broken down into simpler substances. Lavoisier identified 33 elements as pure and indivisible. Of those 33, 20 (of the 112 elements known today) are still considered pure elements. Unfortunately, though a brilliant scientist, he was also associated with French taxation and the ruling governmental class. In 1794, his research was cut short by the guillotine and the French Revolution.

> An **element** is made up of a pure sample with all of the same kinds of atoms and cannot be further separated into simpler components.

Chemical Nomenclature

As time went on, scientists began to study individual elements, but there was a problem. Because chemical elements had different names in different languages, scientists couldn't be sure they were talking about the same thing. Just as a traveler, not knowing the language, has problems in a foreign country asking directions or finding a hotel, early chemists had problems comparing results and analyzing compounds that no one seemed to recognize. Scientists were too busy doing experiments to study languages and ran into trouble when they tried to communicate their findings to colleagues in others parts of the world. For example, the element iron is called *eisen* in German, *piombo* in Italian, *olovo* in Czech, and *fer* in French. To give you a better idea of the problem, Table 3-1 provides a sampling of element names in different languages.

Symbols

Before standard element symbols and the Periodic Table were created, scientists used names for elements based on their root languages. Some element names

TABLE 3-1 Elements have many interesting names in different languages					
Symbol	**English**	**German**	**Swedish**	**French**	**Spanish**
Au	Gold	Gold	Guld	Or	Oro
Bi	Bismuth	Wismut	Vismut	Bismuth	Bismuto
C	Carbon	Kolenlstoff	Kol	Carbone	Carbono
Co	Cobalt	Kolbalt	Kolbolt	Cobalt	Cobalto
Cu	Copper	Kupfer	Koppar	Cuivre	Cobre
Fe	Iron	Eisen	Jarn	Fer	Hierro
Hg	Mercury	Quecksilber	Kvicksilver	Mercure	Mercurio
N	Nitrogen	Stickstoff	Kvave	Azote	Nitrogeno
Na	Sodium	Natrium	Natrium	Sodium	Sodio
P	Phosphorus	Phosphor	Fosfor	Phosphore	Fosforo
Pb	Lead	Blei	Bly	Plomb	Plomo
S	Sulfur	Schwefel	Svavel	Soufre	Azufre
Se	Selenium	Selen	Selen	Selenium	Selenio
Sn	Tin	Zinn	Tenn	Etain	Estano
W	Tungsten	Wolfram	Volfram	Tungstene	Wolframio
Z	Zinc	Zink	Zink	Zinc	Zinc

originated from seldom-used languages like Old English and Latin. For example, the Latin name for copper is *cuprum*, which is why the symbol for copper became Cu.

Common nicknames added to the confusion. For example, baking soda is used to make breads rise, but its chemical name is sodium bicarbonate. Battery acid is the liquid that allows electricity to be mysteriously generated in cars. To a chemist, it's sulfuric acid. Household laundry bleach is known as sodium hypochlorite to chemists. Salt is sodium chloride. Do you know anyone who says, "Please pass the sodium chloride" at dinner?

Unfortunately, common names for elements aren't always helpful since what is common in one geographical region may be unknown in another. In fact, there are too many element combinations for common names to ever be practical worldwide. Many would change with the speaker's culture and references.

> **Chemical nomenclature** is the standardized system used to name chemical elements and compounds.

As chemistry advanced, it became obvious that some sort of common code or *chemical nomenclature* was needed. The symbol for an element can be one letter as in carbon (C) and phosphorus (P), two letters as in seaborgium (Sg) and molybdenum (Mo), or three letters as in the more recent elements in the Periodic Table such as ununbium (Uub) and ununquadium (Uuq).

> When an element has a name with more than a one letter (e.g., zirconium—Zr; meitnerium—Mt), only the **first** letter of the symbol name is **capitalized**.

Element symbols provide a simple shorthand code when writing the full chemical name is impractical. The powerful insecticide, *dichlorodiphenyltrichloroethane* or DDT, is written as $(C_6H_4Cl)_2CHCCl_3$. If it weren't for the carbon (C), hydrogen (H), and chlorine (Cl) symbols, so much time would be taken writing the full name there wouldn't be time for experiments! As you'll see in later chapters, chemical symbols become especially important when writing chemical reactions.

Atomic Number

Atomic number is written as the superscript of an element on the Periodic Table, while the atomic weight is written as a subscript.

> **Atomic number** (Z) equals the number of protons in the nucleus of an atom.

Atomic Weight

The atomic weight of an element is the average of the atomic masses of its natural isotopes. Atomic number is not to be confused with *atomic weight*. The atomic number of phosphorus is 15, while its atomic weight is 30.97.

In 1864, *Die Modernen Theorien der Chemie* (the Modern Theory of Chemistry) was published by German chemist, Lothar Meyer. Meyer used the atomic weight of elements to arrange 28 elements into 6 families with similar chemical and physical characteristics. Sometimes elements seemed to skip a predicted weight. When this happened and Meyer had questions, he left spaces for possible elements. Meyer also used the word, *valence*, to describe the number equal to the combining power of one element with the atoms of another element. He thought this combination of characteristic grouping and valence was responsible for the connection between elements.

After seeing a recurring pattern of peaks and valleys when plotting atomic weight, Meyer thought these patterns formed family rhythms. Because he measured the volume of one atomic weight's worth of an element with the same number of atoms in the sample, Meyer thought the measurements must stand for the same amounts of each distinct atom.

Figure 3-1 illustrates Meyer's experimental results with a recurring pattern in one family. If you start with the element at the top of each peak, you find

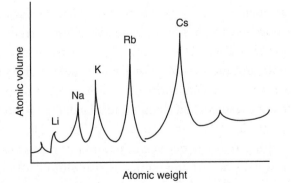

FIGURE 3-1 · Meyer's research results were divided by atomic weight.

lithium, sodium, potassium, rubidium, and cesium lined up by atomic number and weight.

The Octave Rule

Around this same time, English chemist John Newlands was also overlapping elements and seeing similarities. Newlands was a jazzy kind of guy, who noticed chemical groups repeated every eight elements. This pattern reminded him of eight-note music intervals, and so he called his findings, the *Octave Rule*. Members of the Royal Chemical Society were not into music composition and unfortunately ignored Newlands' work for many years.

Unlike Newlands' octaves, Meyer's data showed the groups were not the same length. He was one of the first chemists to notice group lengths started small and got larger. Meyer saw repeating periods of atomic volume. The first period, contained only hydrogen and was one period in length. The second and third periods had seven elements. The fourth and fifth periods were 17 elements in length. Meyer's work held until the inert gases were discovered, then an extra element was added to each period, to give periods of 2, 8, 8, 18, 18. Check out a Periodic Table to see this pattern.

Five years later, working separately, Russian chemist Dimitri Mendeleyev and Lothar Meyer, arranged the elements into seven columns relative to the elements' known physical and chemical properties. There were small differences in their work, but each added to the current knowledge.

Mendeleyev presented a scientifically significant table of elements, *On the Relation of the Properties to the Atomic Weights of the Elements*, which the Russian Chemical Society praised. In his paper, Mendeleyev discussed the periodic similarities between chemical groups with similar reactions.

As it turns out, many of the gaps Mendeleyev had in his periodic chart turned out to be correct placeholders for elements discovered later. Initially, titanium (Ti) was placed next to calcium (Ca), but this would have placed it in group (III) with aluminum (Al). However, because studies of titanium's properties showed that it was more like silicon (Si) than aluminum, Mendeleyev left a space next to calcium and placed Ti in group (IV) with silicon. Ten years later in 1879, Swedish chemist Lars Nilson, discovered the missing element with properties between calcium (atomic weight of 40) and titanium (atomic weight of 48) and named it scandium (Sc). Scandium has an atomic weight of 45. Figure 3-2 shows the element titanium and some of its characteristics.

Ti

Titanium

Atomic number – 22

Atomic mass – 47.90

Group – 4

Period – 4

Transition metal

Electrons per orbital layer – 2,8,10,2

Valence electrons – 1s2 2s2p6 3s2p6d2 4s2

FIGURE 3-2 · Specifics of the metal titanium include its group and period.

In 1870, Meyer's updated Periodic Table of 57 elements was published. This table included properties such as *melting point,* and added depth to the understanding of interactions and the role of atomic weight. Meyer also studied elements' atomic volume to help pin down the element placement into specific groups.

Perhaps Meyer's curiosity came from the fact that he grew up in a family of physicians and was exposed to scientific and medical discussions early in his life. His initial schooling in Switzerland was in the field of medicine. The many elements in the body and their complex interactions gave Meyer much to think about. Table 3-2 shows a few of the most common elements and their functions in the body.

Meyer's initial chemistry research grew out of his fascination with the physiology of respiration. He was one of the first scientists to recognize that oxygen bonded with hemoglobin in the blood. Then, in order to explain specific

TABLE 3-2 Elements serve many important functions in the body

Element	Functions in the body
Calcium	Bones, teeth, and body fluids
Phosphorus	Bones and teeth
Magnesium	Bone and body fluids, energy
Sodium	Cellular fluids, transmission of nerve impulses
Chloride	Dissolved salt in extracellular and stomach fluids
Potassium	Cellular fluids and transmission of nerve impulses
Sulfur	Amino acids and proteins
Iron	Blood hemoglobin, muscles, and stored in organs

biochemical processes and complex systems, he had to identify the elements more completely.

The modern Periodic Table contains around 118 elements. Those up through atomic number 92 (uranium) are naturally occurring, whereas the "transuranic" elements, those synthesized in heavy nuclei interactions, make up the most recent discoveries. Some symbols of elements, like in Meyer and Mendeleyev's time, represent gaps for any elements that seem to be hinted at by test data. When compared to Mendeleyev and Meyer's early tables, the details described over 150 years ago are amazingly accurate. A commonly used Periodic Table is shown in Fig. 3-3.

Periodic Table of Elements

<u>Periodic Table History</u>

	I A																	VIII B
1	1 **H** 1.0079	II A											III B	IV B	V B	VI B	VII B	2 **He** 4.003
2	3 **Li** 6.94	4 **Be** 9.0121											5 **B** 10.81	6 **C** 12.011	7 **N** 14.006	8 **O** 15.999	9 **F** 18.998	10 **Ne** 20.17
3	11 **Na** 22.989	12 **Mg** 24.035	III A	IV A	V A	VI A	VII A	VIII A	VIII A	VIII A	I B	II B	13 **Al** 26.981	14 **Si** 28.085	15 **P** 30.973	16 **S** 32.06	17 **Cl** 35.453	18 **Ar** 39.948
4	19 **K** 39.098	20 **Ca** 40.08	21 **Sc** 44.955	22 **Ti** 47.90	23 **V** 50.941	24 **Cr** 51.996	25 **Mn** 54.938	26 **Fe** 55.847	27 **Co** 58.933	28 **Ni** 58.71	29 **Cu** 63.546	30 **Zn** 65.38	31 **Ga** 69.735	32 **Ge** 72.59	33 **As** 74.921	34 **Se** 78.96	35 **Br** 79.904	36 **Kr** 83.80
5	37 **Rb** 85.467	38 **Sr** 87.62	39 **Y** 88.905	40 **Zr** 91.22	41 **Nb** 92.906	42 **Mo** 95.94	43 **Tc** 98.906	44 **Ru** 101.07	45 **Rh** 102.90	46 **Pd** 106.4	47 **Ag** 107.86	48 **Cd** 112.41	49 **In** 114.82	50 **Sn** 118.69	51 **Sb** 121.75	52 **Te** 127.60	53 **I** 126.90	54 **Xe** 131.30
6	55 **Cs** 132.90	56 **Ba** 137.33	57 **La** 138.90	72 **Hf** 178.49	73 **Ta** 180.94	74 **W** 183.85	75 **Re** 186.20	76 **Os** 190.2	77 **Ir** 192.22	78 **Pt** 195.09	79 **Au** 196.96	80 **Hg** 200.59	81 **Tl** 204.37	82 **Pb** 207.2	83 **Bi** 208.98	84 **Po** (209)	85 **At** (210)	86 **Rn** (222)
7	87 **Fr** (223)	88 **Ra** 226.02	89 **Ac** (227)	104 **Unq** (261)	105 **Unp** (262)	106 **Unh** (263)	107 **Uns** (262)	108 **Uno** (265)	109 **Une** (266)	110 **Unn** (272)								

Lanthanide Series	58 **Ce** 140.12	59 **Pr** 140.90	60 **Nd** 144.24	61 **Pm** (145)	62 **Sm** 150.4	63 **Eu** 151.96	64 **Gd** 157.25	65 **Tb** 158.92	66 **Dy** 162.5	67 **Ho** 164.93	68 **Er** 167.26	69 **Tm** 168.93	70 **Yb** 173.04	71 **Lu** 174.96
Actinide Series	90 **Th** 232.03	91 **Pa** 231.03	92 **U** 238.02	93 **Np** 237.04	94 **Pu** (244)	95 **Am** (243)	96 **Cm** (247)	97 **Bk** (247)	98 **Cf** (251)	99 **Es** (254)	100 **Fm** (257)	101 **Md** (258)	102 **No** (259)	103 **Lr** (260)

Color Reference for element type:	
Noble Gas	Halogen
Metal	Rare Earth
Trans. Metal	Non Metal
Alkali Metal	Alkali Earth

FIGURE 3-3 • The Periodic Table is an important tool for the chemist.

Much more of the very latest information can be found on current and suspected elements at a variety of interactive Internet sites including: http://www.webelements.com and http://www.ptable.com/.

Though Thompson, Rutherford, Meyer, and Mendeleyev didn't quite understand what caused many of the reactions they saw, they recorded the elemental patterns anyway. They noticed some elements had properties reflecting atomic weight and atomic number, but weren't really sure why. Modern chemistry has since found answers to these questions.

Classes of Elements

Like a family history, elements are arranged in family groups such as noble gas, halogen, metal, rare earth, transition metal, nonmetal, alkali metal, and alkaline earth. Just as genetics helps biologists and physicians find a person's make-up, so the grouping of elements into families and groups makes it possible to see related properties for different elements. As early scientists discovered, some elements with similar properties can be grouped together. Basically, groups of elements are divided into four main classes:

1. Representative elements (groups IA – VIIA)
2. Noble or inert gases (group VIII)
3. Transition metals (group B elements)
4. Inner transition metals (lanthanide and actinide series)

The *representative elements* or main group of elements is further divided into groups with characteristic properties. Group IA, the *alkali metals* (e.g., lithium, potassium), are all soft metals (except hydrogen, a gas) that react violently with water. Group IIA, the *alkaline earth metals* (e.g., beryllium, strontium), are also chemically reactive. The *chalcogens*, group VIA, (e.g., oxygen, sulfur) are named from the Greek *chalkos* meaning ore. Many ores are made with varying amounts of oxygen and sulfur. The rest of the representative element groups (IIIA through VA) have not been given descriptive names. The *halogens* in group VIIA are all nonmetals. The *noble* or *inert gases* (group VIII) get their name from their lack of reactivity or inertness. These elements are usually too unsociable to form chemical compounds. In fact, helium, neon, and argon refuse to play with anyone and don't form any compounds at all. All these gases exist naturally as individual atoms in the environment.

Periods and Groups

The Periodic Table of the elements is the most important tool of general chemistry. Probably only the Bunsen burner (laboratory gas flame) rivals it. The amount of information pulled together in one place makes calculations, reactions, and the study of matter a whole lot easier to decipher. Elements with similar boiling or melting points usually act the same way when exposed to the same experimental conditions. The same is true of freezing and vaporization points.

The Periodic Table is divided into rows and columns, known as, *periods* and *groups*. Dividing elements into periods and groups helps classify them by their specific characteristics.

Periods

There are seven periods in the Periodic Table. Each period (rows starting on the left-hand side of the Periodic Table) ends with an element known as a *noble gas*. Like kings and queens in impenetrable castles, these gases are chemically unreactive and composed of individual atoms.

> A **period** contains chemical elements in a horizontal row of the Periodic Table.

When studying the elements in each period, it is clear the number of elements increases with each period, which will be explained in Chap. 7. For now, just know the first period contains only hydrogen (H) and helium (He). The second period has eight elements lithium (Li) through neon (Ne). The third period also contains eight elements, sodium (Na) through argon (Ar). The fourth period has 18 elements, potassium (K) through krypton (Kr). The fifth period also has 18 elements, rubidium (Rb) through xenon (Xe). The sixth period has 32 elements, cesium (Cs) through radon (Rn).

To make the table more manageable, the sixth and seventh periods have been divided between 57, 58 and 89, 90. These rows are shown fully expanded at the bottom of the chart. The seventh period is not complete but has placeholders just as Meyer had for the later discovery of elements. The most recently named elements, roentgenium (Rg, #111) and copernicium (Cn, #112) are both thought to be in the seventh period and metallic in nature. Debated element placeholders up to element #118 are not included on older Periodic Tables.

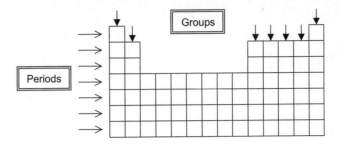

FIGURE 3-4 · Periods and groups of elements are found in columns and rows of the Periodic Table.

Groups

The groups (columns starting on the left-hand side of the Periodic Table) are numbered most frequently with Roman numerals. The *International Union of Pure and Applied Chemistry* (IUPAC), in order to avoid confusion, set up a standard numbering plan in which columns were numbered I through VIII, according to their characteristics.

> A **group** contains the elements in one column of the Periodic Table.

These groups are further divided into A and B subgroups with the A groups called the main groups or *representative* elements and the B groups called the *transition* elements. Numbers 58 (cerium) through 71 (lutetium) are known as the *Lanthanide* series and 90 (thorium) through 103 (lawrencium) as the *Actinide* series of elements. The Periodic Table divided into periods and groups is shown in Fig. 3-4.

Metallurgy—The Chemistry of Metals

> **Metallurgy** describes the science of metals and their properties, as well as purification methods, reactions, and the formation of useful alloys.

Metals versus Nonmetals

Metals and nonmetals, commonly designated by a heavy zigzag line on the Periodic Table with metals to the left and nonmetals to the right have different properties. Figure 3-5 shows this dividing line.

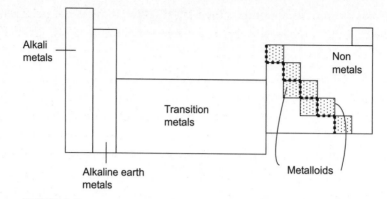

FIGURE 3-5 · Metalloids are found at the dividing line between metals and nonmetals.

Metals, about 80% of the elements, can be pulled into thin wires (i.e., ductile) or pounded into sheets (i.e., malleable). They are usually solid at room temperature, shiny, and good conductors of heat and electricity. Mercury is the only metallic element that is liquid at room temperature.

Nonmetals are gases at room temperature (e.g., helium and argon) or brittle solids (e.g., phosphorus and selenium). Bromine is the only element in the nonmetal group that is liquid at room temperature.

Alkali Metals

Within the main group of elements (Group A) are many different types. Within the *alkali metals* (IA group), lithium, sodium, and potassium are known as active metals. Their reactivity increases as their atomic number increases. These elements are extremely reactive in water and air. They are frequently stored in oil to prevent explosions when accidentally mixed with water.

Although all highly reactive, alkali metals have various properties. Lithium is the third element in the Periodic Table (IA group) and is usually found in the mineral spodumene. Lithium is the lightest metal and is so soft when isolated it can be cut with a sharp knife. Lithium's density is so low it could float on water, but like the rest of the alkali metals, it is extremely reactive in water. Generally, lithium is stored in oil or kerosene. When combined with aluminum, lithium forms a strong, lightweight alloy metal used in aircraft and spaceships. Similarly, sodium and potassium are soft alkali metals, but as more brittle salts, they form silicate minerals found in seawater.

Like the rest of the alkali metals, cesium is silvery white in purified form. It is commonly found in the mineral pollucite, a compound containing silicon, oxygen, and aluminum. Cesium is the softest metal known and melts at 28°C. When held, it will melt in your hand at body temperature, 37°C (98.6°F) like a piece of chocolate.

Alkaline Earth Metals

The Periodic Table places elements according to their specific characteristics. An element's location provides information about its "personality" and uses.

Barium's location (IIA) shows that it falls into a group of elements called alkaline earth metals. Barium has medical applications in the form of barium sulfate (e.g., opaque to x-rays and used to image the digestive tract) and in photography (e.g., whitener in photographic papers). Calcium and magnesium are also located in the IIA group of the Periodic Table and minerals in the diet.

Despite similarities, these elements are different. For example, barium (3.63 g/cm²) has over twice the density of calcium (1.54 g/cm²). Conversely, barium [0.18 watts/(cm K)] has only a fraction of calcium's thermal conductivity [2.01 watts/(cm K)].

Transition Metals

The *transition metals* (Group B) contain the many recognizable metals. They are used in construction, coins, and jewelry. The transition metals group includes iron, nickel, and chromium.

Gold, Silver, and Copper Gold, silver, and copper are sometimes called *noble metals* because they were historically the metal of kings and nobles. Since the first shiny speck caught the eye of early humans, gold, silver and copper have been used for coins, jewelry, and household serving ware. Resistant to rust and corrosion, they were an excellent choice for anything meant to last forever (e.g., coins).

Gold is a shiny yellow metal that conducts heat and electricity well. It is the most malleable and ductile metal. The early alchemists based their reputations and lives on creating gold for their patrons. In the western United States of the 1800s, gold fever affected thousands of people during the gold rush days who were seeking their fortunes and a better life.

Silver is a brilliant white, lustrous metal that is the best conductor of heat and electricity of all the metals. It was also prized by early people for its beauty and uses. Unlike gold, silver is less resistant to corrosion and will tarnish, turning black upon oxidizing in the air. According to historians, the state of Nevada was admitted to the Union in 1864, during the Civil War to provide funds to the Union through easier access to its silver resources. Silver is used in coins, jewelry, electrical contacts, mirrors, circuitry, photography, and batteries.

Copper is orange-brown in color and used in pipes, electrical wires, coins, paints, fungicides, and alloys combined with other metals. In many countries, local artisans use copper widely to make platters, bowls, tools, and jewelry. Pennies in the United States were once 100% copper, but are now (since 1981) only plated on the outside with copper for a reddish-brown (copper) color. Many years ago, policemen's badges were made of copper and so officers were tagged with the nickname "copper" or "cop" for short.

Iron Iron is a commonly known metal discovered thousands of years ago and frequently used as a progress marker of human civilization. Iron made a huge difference in the way people hunted, cooked, and fought one another. Because of the importance of these changes, the time period up until about 1100 B.C. is known as the Iron Age.

Iron is a reddish-brown "friendly" metal that is safe to handle and readily combines with many other elements to form products known for their strength. Since primitive times, objects such as hand tools, cups, and plows were made from iron. Iron is the fourth most plentiful element in the Earth's crust making up about 5% of the elements present after oxygen, silicon, and aluminum, respectively. Currently, iron makes up more than 90% of all refined metal worldwide.

Iron is found naturally in several ores, the main one being *hematite* (Fe_2O_3). Hematite has different colors and forms. The silvery black ore is used for jewelry. Red hematite is used as a paint pigment known as red ocher.

Inner Transition Metals

The *inner transition metals* are made up of a group of 15 *rare earth* metals or *lanthanide series*. These silvery-white metals are very similar to each other in their chemical properties and are found in permanent magnets and headphones.

Another group of inner transition metals, the *actinide series*, (e.g., uranium, americium, and neptunium) are mostly human-made elements. The actinide series are difficult to isolate, because they are highly unstable and undergo radioactive decay. For example, uranium has compounds in each of the +3, +4, +5, and +6 oxidation states. They are radioactive and used in advanced smoke detectors, neutron-detection devices, and in nuclear reactions. We will learn more about oxidation in Chap. 10.

Metallic Crystals

Metals form large crystalline structures with high boiling and melting points. These structures are made of metal ions. Electrons associated with metal atoms move around within crystalline structures and cause the solid crystals to be good conductors, because electricity is all about moving electrons!

You can think of these structures as a vegetable soup with the "broth" made up of the electrons and the "vegetables" of interconnected, positively charged metal ions. The broth does not have enough electrons to form individual bonds between atoms, so electrons are shared for efficiency. This strong bonding makes the metallic crystal harder, since ions with shared electrons are held more tightly between atoms.

Nanocrystals   *Nanocrystals* are clumps of atoms (e.g., few hundred to thousands of atoms) that combine into a crystalline form of matter. Commonly around 10 nm in diameter, nanocrystals are larger than molecules, but smaller than bulk solids and often have physical and chemical properties somewhere in between. Because a nanocrystal is pretty much all surface and no interior, its properties often change significantly with increases in size.

One of the big advantages of nanocrystals over larger materials is that their size and surface can be precisely controlled. You can change how a nanocrystal conducts charge, its crystalline structure, and even its melting point.

Some spherical nanocrystals have a shell of cadmium sulfide and core of cadmium selenide. Different-sized nanocrystals are useful for emitting multiple colors of light. These nanocrystals have lots of possible applications, including use as super sensitive fluorescent labels for studying biological materials. In fluorescent labeling, markers (e.g., antibodies attached to specific proteins) are tagged with dye molecules that fluoresce or emit a particular color of light when stimulated by photons.

Quantum Dots   *Quantum dots* are metal oxide crystals that function like artificial atoms with special electronic properties. They are charged and give off different colors of light depending on size and specific energy levels. These energy levels can be limited by changing the size and shape of the quantum dot. They don't always transfer or "conduct" electrons well, so quantum dots are classified as *semi*conductors.

> A **quantum dot** (i.e., nanodot) is a semiconductor nanocrystal that is a few nano-meters to a few hundred nanometers in overall size.

Like atoms, quantum dot energy levels can be studied with optical spectroscopy methods.

Because of their optical features, different-sized quantum dots (with different color/wavelength absorptions) can be seen in solution with the naked eye.

Electron microscopes use electrons, instead of light, to view super-small objects. The wave-like action of tiny particles is a lot like light. The term *de Broglie wavelength* describes the wavelength of these particles. Since electrons commonly have more forward motion than photons, their de Broglie wavelength is smaller, which allows sharper spatial resolution.

A **de Broglie wavelength** is the measure of wave movement (wavelength) of a particle. The wavelength (λ) is given by $\lambda = h/mv$ (where h is the Planck's constant, m is the particle mass, and v is velocity).

Even though quantum dot composition is important, size establishes color. As a particle's size increases, the fluorescence becomes redder (i.e., less energy) in the color spectrum. Conversely, the smaller the size, the fluorescence becomes bluer (higher energy) in the color spectrum. Some researchers suggest shape may also be a factor in quantum dot colorization, but more testing is needed.

It is also possible to tune a quantum dot. Size impacts other properties as well. The larger and more red-shifted a quantum dot becomes, the less quantum properties are involved. The smaller a quantum dot, the easier it is to take advantage of unique quantum properties.

Researchers use quantum dots that have been absorbed through a cell membrane to tag surfaces within the cell. Additionally, quantum dots don't "bleach" or lose their color intensity, which makes medical images clearer. Medical clinicians are better able to study the inner workings of biological systems through quantum dot fluorescence.

Alloys

Other metals, such as chromium, manganese, nickel, tungsten, cobalt, and chromium combine with iron to make steel, an *alloy* of superior strength, hardness, and durability. About 99% of all iron mined in the world today goes into the manufacture of steel.

Alloys are made from the combination of two or more metals or a metal and nonmetal.

When iron is coated with zinc, it is resistant to the rusting effects of oxygen and is said to be *galvanized*. Nails, wire, and large sheets of metal are treated this way to increase the life and usefulness of such metals.

By mixing metals of different characteristics in different proportions, a new metal or *metal alloy* can be created. The metals are melted together into a molten solution, cooled and allowed to become solid again. The newly formed solid has the characteristics of the parent metals along with new properties, such as greater strength than either of its parents.

EXAMPLE 3-1

Steel = 80% iron, 12% chromium, 8% nickel. However, its cousin, a manganese steel alloy = 87% iron, 12% manganese, 1% carbon. The alloy used in railroad rails is much harder.

Iron alloys are made by mixing together two or more molten metals and a nonmetal. When producing 10-carat, 14-carat, and 18-carat gold, various percentages of gold, copper, and silver are used. Table 3-3 gives an idea of commonly used alloys and their composition.

Mercury is prized by scientists for its ability to dissolve other metals and form alloys. When combined with silver, dentists make a silver-mercury *amalgam* (i.e., alloy) to fill cavities in teeth. Mercury also dissolves gold and is used to extract gold from other ores.

TABLE 3-3 When metals are mixed in set quantities, alloys of specific properties are made

Alloy	Composition
Brass	72% copper, 28% zinc
Bronze	93% copper, 7% tin
Carbon steel	1% manganese, 0.9% carbon, 98% iron
Manganese steel	12% manganese, 1% carbon, 87% iron
Stainless steel	12% chromium, 8% nickel, 80% iron
Chromium steel	3.5% chromium, 0.9% carbon, 95.6% iron
Sterling silver	92.5% silver, 7.5% copper
10-carat gold	42% gold, 12% silver, 40% copper
14-carat gold	58% gold, 24% silver, 17% copper, 1% zinc
18-carat gold	75% gold, 18% silver, 7% copper
24-carat gold	100% gold

Metalloids

The elements found on the border between metals and nonmetals are known as *semimetals* or *metalloids* since they have the characteristics of both metals and nonmetals. Boron, silicon, germanium, arsenic, antimony, tellurium, and polonium are all metalloids. (Refer back to Fig. 3-5.)

> **Metalloids** are found on the zig-zag borderline of the Periodic Table between metals and nonmetals.

Silicon, the second most abundant element in the Earth's crust after oxygen, is found in granite, quartz, clay, and sand. In its most basic form, silicon's crystalline structure is much like that of a diamond. Maybe that is why when you walk along the beach and the sun is high, grains of sand glisten like diamonds.

In semiconductors, silicon is used to carry an electrical charge. Silicon's natural conducting capacity is augmented when small amounts of arsenic or boron are added to it. Silicon is also used to make lubricants, computer circuits, and medical implants and joints.

Nonmetals

Nonmetals are found at the far right side of the Periodic Table and consist of 18 elements (counting from hydrogen with the nonmetals to the right). Six of these elements are the noble gases in group VIII in the far right column.

Nonmetals are described as elements without metallic properties such as ductility or malleability. They are generally gaseous or solid and serve as poor thermal or electricity conductors. Some nonmetals, like diamonds (a stable crystalline form of carbon) are very hard and unreactive, while others, like chorine, have free electrons and are more reactive. Nonmetals are not lustrous, malleable, or ductile. Eleven of the nonmetals are gases, one is a liquid, and six are solids. In Fig. 3-6 nonmetals are grouped into solid, liquid, and gaseous forms.

Inert gases are nonmetals and exist as individual atoms, such as neon and argon. Other nonmetals are two-atom molecules, such as hydrogen, nitrogen, and oxygen.

Halogens

Nonmetals located in column (VIIA) of the Periodic Table are known as *halogens* or "salt formers." They easily accept electrons from other atoms and

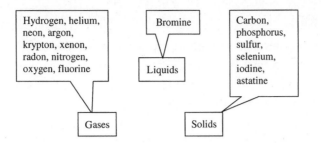

FIGURE 3-6 · Nonmetals are found in all three forms of matter, that is, liquid, solid, and gas.

combine with metals to form salts. Chlorine, fluorine, bromine, iodine, and astatine are in the halogen group.

In nature, fluorine is found in ores of fluorspar (e.g., calcium fluoride) and cryolite (i.e., a combination of sodium, fluorine, and aluminum). In the body, fluorine is found in the blood, bones, and teeth. Many communities in the United States add small amounts of fluorine to drinking water, because fluorine can strengthen teeth and prevent cavity formation.

Naming Metals and Nonmetals

In general, metals form positive ions (i.e., cations), while nonmetals form negative ions (i.e., anions). Remember noble gases are independent and don't care to form cations or anions.

When a metal cation and nonmetal anion form a compound, they are called "metal-nonmetal *binary compounds*." Metals with ions of only $^+1$, $^+2$, or $^+3$ charges in groups IA and II through III are called *monoatomic* and *polyatomic* ions, respectively. Lithium, potassium, and cesium all have a $^+1$ charge. Beryllium and strontium, in group IIA are named by the same rules.

However, when you name elements in groups IVA through VIIA, where anions of $^+1$, $^-1$, or $^-3$ anions are formed, then "-ide" is added to the name.

EXAMPLE 3-2

When writing the name for K^+Cl^-, the metal is written first, *potassium*, followed by the nonmetal, *chlorine*. So it becomes *potassium chloride*, since the combined potassium cation ($^+1$) and chlorine anion ($^-1$) use the rule of adding "-ide" to the nonmetal.

EXAMPLE 3-3

To write the name of $Co_2^+S_2^-$, you first name the metal, *cobalt*, followed by the nonmetal, *sulfur* to get *cobalt sulfide*.

Polyatomic Ions

When naming a compound containing two nonmetals, the same rule applies. Write the nonmetal acting like a metal first, followed by the nonmetal second. Usually, the quasi-metal (nonmetal #1) is close to the zigzag border between metals and nonmetals on the Periodic Table.

Carbon, for example, is written first in carbon dioxide (CO_2). Hydrogen is always written first in compounds such as hydrogen fluoride (H^+F^-).

When several atoms make up a compound, Greek prefixes are used as shown in Table 1-4. Examples of this are *di*nitrogen trioxide (N_2O_3) and carbon *tetra*-fluoride (CF_4). More about ions and their properties will be described in Chap. 9.

If you are asked to learn only one thing in all of chemistry, pick the Periodic Table. Learn it and most everything else will fall into place.

QUIZ

1. In 1864, *Die Modernen Theorien der Chemie* was written by which one of the following?

   A. Johnnes Kepler
   B. Natasha Alvandi
   C. Lothar Meyer
   D. Dimitri Mendeleyev

2. The modern Periodic Table contains how many named elements?

   A. 33
   B. 58
   C. 112
   D. 123

3. The following are all names for the element, iron, except

   A. *jarn.*
   B. *etain.*
   C. *eisen.*
   D. *hierro.*

4. Which of the following elements is a gas at room temperature?

   A. Iridium
   B. Antimony
   C. Gallium
   D. Xenon

5. Period 6 of the Periodic Table, contains all of the following elements, except

   A. rhodium.
   B. tantalum.
   C. cesium.
   D. thallium.

6. Which of the following element groups have the most elements?

   A. Metalloids
   B. Nonmetals
   C. Transition/inner transition metals
   D. Actinides

7. Which element is the best conductor of electricity?

   A. Silver
   B. Gold
   C. Mercury
   D. Copper

8. **Which alloy is often used to fill dental cavities?**
   A. Iron
   B. Strontium
   C. Mercury
   D. Phosphorus

9. **Which group of elements is often stored in oil to prevent explosions?**
   A. IA
   B. IIA
   C. IB
   D. IIB

10. **Sg is the chemical symbol for which element?**
    A. Sodium
    B. Samarium
    C. Scanium
    D. Seaborgium

chapter 4

Solids and Liquids

Early cultures thought matter was everything seen and measured, like salt, grain, and olive oil. Everyone used their senses to define what they saw, heard, tasted, touched, and smelled. Anything that could not be sensed did not exist or belonged in the realm of myths and legends. People only believed in what they saw "with their own eyes."

CHAPTER OBJECTIVES

In this chapter, you will

- Learn the main types of solids
- Find out important differences between solids and liquids
- Discover the four different types of bonding in crystalline solids
- Learn about the properties of liquids (e.g., condensation, boiling point)
- Find out why chemists use a phase diagram when studying samples under different pressures and temperatures

What Are Solids?

Solids include things like boulders, metals, crystals, and glass. They are fixed in shape and rigid with a measurable volume. For a solid to change shape in a major way, a strong outside force like fire or heavy impact is needed.

Solids are unyielding. They are made to last. When archeologists excavate ancient sites, solids are still there, but liquids and gases are long gone. We base 90% of what we know of past civilizations on solids left behind, such as fragments of pottery, metal tools, and weapons. But why are some types of matter members of the solid club and not others?

The denser the atoms or molecules in a sample, the more tightly packed and solid it is. Figure 4-1 shows the main difference between solids and other forms of matter is density. Solids are denser than liquids or gases, but there is a lot more to it than that. Solids have a lot of quirks that make the club fairly interesting.

Amorphous Solids

Solids are normally either *amorphous* or *crystalline*. The first group, amorphous is made up of shifting members that can't decide whether to be solid or not. These are things like wax, rubber, glass, and polyethylene plastic.

> **Amorphous solids** have no specific form or standard internal structure.

Brittle, noncrystalline solids tend to shatter into sharp pieces when broken. These amorphous solids are less dense than other crystalline club members and have no definite melting point. When heated, they slowly soften and become very

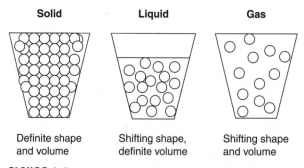

Solid	Liquid	Gas
Definite shape and volume	Shifting shape, definite volume	Shifting shape and volume

FIGURE 4-1 · The three forms of matter consist of solids, liquids, and gases.

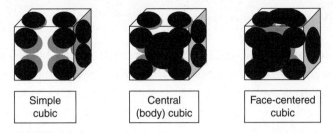

| Simple cubic | Central (body) cubic | Face-centered cubic |

FIGURE 4-2 • The beauty of crystalline solids comes mainly from their atomic arrangements.

flexible. Have you ever heated glass tubing in a flame to make a 90° angle in a straight rod? This is an example of the changing shape of noncrystalline solids.

Crystalline Solids

Most people are a lot more familiar with crystalline solids. These solids include quartz, diamond, salt, and different gemstones. The atoms or molecules of crystalline solids form specific crystal patterns of an ordered lattice or framework, a lot like stacked bricks.

Crystalline solids are arranged into regular shapes based on a cube (i.e., simple, central, and face-centered). Figure 4-2 shows three ways the molecules of a crystalline solid can be arranged. As a molecule goes from a simple cubic structure to a face-centered cubic structure, density increases. There is less space between the atoms, so it is more tightly packed. Higher density makes materials harder and less flexible.

Unlike amorphous solids, a lattice structure provides predictable breaks along set lines. This is why diamonds and gemstones can be cut into facets. The round, oval, pear, emerald, and diamond-shaped cuts used in jewelry can be created by different gem cutters all over the world due to their characteristic lattice structures.

Crystallization and Bonding

There are four different types of bonding in crystalline solids. These determine whether a solid will be molecular, metallic, ionic, or covalent.

Molecular Solids

These crystalline solids have molecules at the corners of the lattice instead of individual ions. They are softer, less reactive, have weaker nonpolar ion attractions and lower melting points.

A molecular solid is held together by intermolecular forces. These forces are further described in Chap. 8. Because these forces are weaker than covalent or ionic bonds, molecular solids are soft and have fairly low melting points, density, and hardness. Hydrocarbons, sugar, and sulfur are examples of molecular solids.

Metallic Solids

Metals make up another type of crystalline solid. The temperature needed to break the bonds between positive metal ions in specific lattice positions, like iron in iron (II) disulfide (FeS_2), and the happy valence electrons around them is fairly high. This strong bonding gives stable molecules flexibility and allows them to be formed into sheets and strands without breaking, as we learned in Chap. 3.

A metallic solid like iron or silver is held together by a network bonding type. A positive core of metallic atoms is held together by a surrounding field of negatively charged electrons. This array of (+) metal ions and diffuse valence electrons (−) makes them good conductors of electricity.

Over 80 elements in the periodic table are metals. Metals are solid at ordinary temperature and pressure, with some exceptions (e.g., mercury and gallium). They have high luster and reflectivity, and can be soft (e.g., cesium) to hard (e.g., tungsten).

Since metals atoms are all the same, they can't form ionic bonds. The force binding metals atoms together is called *metallic bond*.

Ionic Solids

Ionic solids form a lattice so that the outside points are made up of ions instead of larger molecules. These are the "opposites attract" solids. The different forces give these hard, ionic solids (e.g., magnetite and malachite) high melting points and cause them to be brittle. They also don't conduct electricity in the solid state compared to metals. Figure 4-3 shows the melting points of different solids.

Ionic crystals with the ions of two or more elements form three-dimensional crystal structures held together by strong ionic bonds. Ionic bonding in a solid takes place when anions (−) and cations (+) are held together by the electrical pull of opposite charges. Opposites attract! This electrical magnetism is found in a lot of salts like sodium chloride (NaCl), calcium chloride (CaCl) and zinc sulfide (ZnS).

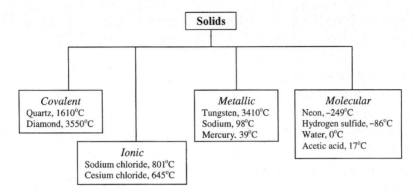

FIGURE 4-3 • Depending on bonding strength, solids have a wide variety of melting points.

For this reason, ionic compounds are called *electrolytes*. Electrolytes carry a current when melted or dissolved in a solvent during electrolysis. Electrolytes include compounds that ionize in water and assist the flow of electricity. Soluble ionic compounds like salts and some minerals are all electrolytes (e.g., potassium chloride).

> An **electrolyte** is a substance with free ions that allows it to dissolve in water and conduct electricity.

? Still Struggling

Some sports drinks advertise the importance of replacing electrolytes in the body after exercising. They refer to body processes that rely upon electrolytes to carry electrical signals responsible for triggering muscle contractions and sending nerve impulses to other cells. Without electrolytes, cells wouldn't be able to communicate well.

Covalent Solids

Covalent crystals are held together by single covalent bonds. This type of stable bonding produces high melting and boiling points. Assembled in large nets or chains, covalent multilayered solids are extremely hard and stable in this type of pattern. However, covalent solids are bad electrical conductors compared to metallic solids.

Diamond atoms use this type of network structure to form three-dimensional solids. In a diamond, a carbon atom is covalently bonded to four other carbons. This strong crystalline structure makes diamond the hardest known carbon-based solid.

Carbon's different bonding arrangements form a diamond (e.g., pyramid-shaped), graphite (e.g., flat-layered sheets), or buckminsterfullerene (e.g., C60, C70, soccer ball-shaped). These examples show the variety and stability of covalent molecules. Nets, chains, and balls of carbon bonded into stable molecules make these solids hard and stable.

Allotropes are different structural forms of the same element. Graphite, diamond, and buckminsterfullerene are all allotropes of carbon.

Properties of a Solid

The unique character or way an element reacts is one of its properties. These properties are grouped into two classes, *physical* and *chemical*.

Physical properties make up a sample's physical composition. Physical properties include: color, form, density, heat/electrical conductivity, and melting/boiling points. These properties can be seen without any change in form or dimension.

For example, solid barium is silvery white in color, melts at 727°C (1341°F) and burns green at a wavelength of 554 nanometers. Even melted, it is still barium, but in a different form than when it was solid.

Chemical properties are characteristics involving a substance's behavior when mixed with another element or compound. For example, when copper is exposed to oxygen, it turns green due to a chemical reaction. The thin, green surface layer is copper carbonate or copper sulfate, which actually protects the surface from further corrosion.

Regular parts of a solid's personality (properties) can be used to classify it. These properties can also be found by an element's placement in the Periodic Table. Metals found in the middle of the Periodic Table share a lot of the same characteristics, like brothers and sisters in the same family often have the same hair or eye color. Some of these properties are hardness and structure, melting point, and electrical conductivity. Table 4-1 lists these personality traits of solids.

TABLE 4-1 The individuality of solids can be found in their variety of characteristics
Solids
Fixed volume
Definite structure
Cannot be compressed
Held together by molecular, ionic, metallic, or covalent bonds
Do not expand when heated
Amorphous or crystalline in form
High density

Density of a Solid

Compared to liquids, the atoms in a solid are tightly bonded and a lot more compressed. Think of them like people packed on a commuter train at rush hour on a Monday morning. They are basically stuck in one spot until some outside force changes to allow them to move more freely. Some of these outside forces are bonding changes or reactions, crystallization, heat, and pressure.

Density determines how materials interact. For example, the density of balsa wood is approximately 0.13 g/cm³ and the density of water (20°C) is 0.998 g/cm³. It makes sense, then, that porous balsa wood floats easily on water since its density is so much less than the density of water.

In the case of two metals, the density of copper is 8.96 g/cm³ compared to the density of mercury at 13.55 g/cm³. If a copper penny is dropped into a container of mercury, would it sink or float? Take a look at their densities. Mercury is much denser, so a copper penny would float on the surface of mercury. If the same penny fell into a beaker of water, what would happen?

Temperature and Solids

One of the major players in the solid club is temperature. Temperature controls a lot of who gets along and whether they play nice.

Look at water, normally a member of the liquid club, but exposed to very low temperatures and pow! Ice. Not only a solid, but see-through as well. Thanks to this special gift of H_2O, we enjoy popsicles, ice hockey, and snow balls.

What happens when we add more heat? A solid reaches its melting point. No more ice. It's back to the liquid club. Without ice, we have skiing on rocks (very hard on skis), no polar ice caps, and soupy ice cream.

Every element can be a solid if the temperature is dropped low enough, even gases. This helps us understand the conditions on planets like Jupiter and Saturn in our solar system. They are mostly made up of frozen gases.

Precipitation of Solids

Precipitation describes the formation of a solid during a chemical reaction in a solution. After the reaction takes place, the solid formed is called a *precipitate*. Precipitates can form when two soluble salts react in a liquid to form one or more insoluble compounds.

> An insoluble solid that falls out of solution is called a **precipitate**.

A solid precipitate can be formed from a chemical reaction or in a *saturated* solution. *Saturation* is the point where the solution of a substance can't dissolve any more of the substance (i.e., no more reactions can occur). Figure 4-4 shows how a solution progresses from its initial soluble state to a suspension. It is separated into a *supernate* (i.e., extra liquid floating above) and precipitate.

Commonly, a solid precipitate falls out of the dissolved state and sinks to the bottom of the solution. (Find out more about solubility in Chap. 6.) Precipitates can also form when a solution's temperature drops. Lower temperatures reduce a salt's solubility causing it to precipitate as a solid.

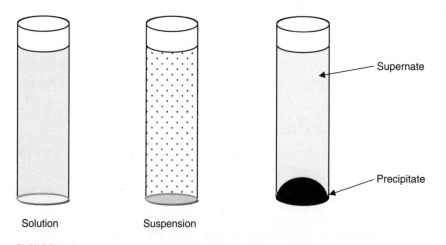

Solution   Suspension

FIGURE 4-4 • Over time, the supernate and precipitate move apart in a solution.

In physical chemistry, saturation describes the point at which a binding site is full. In environmental science, nitrogen saturation refers to how a bionetwork (e.g., soil) has reached it maximum binding ability and is unable to store any more nitrogen.

Mixtures

Mixtures can be separated into two or more substances manually. No chemical reaction is needed. In nature, salt water can be separated back into water and salt by letting the water evaporate.

Mixtures are found in two forms: *heterogenous* and *homogeneous*. A heterogenous mixture is one with physically separate parts and different properties. An everyday example is a mixture of salt and pepper. A heterogenous mixture has separate phases. A *phase* represents the number of different homogeneous materials in a sample. Salt is all one phase and pepper is another phase. They do not have a lot of overlapping characteristics, but are physically separate.

A homogeneous solution has one phase (e.g., liquid) but may have more than one component within a sample. Again, salt water is an example of a homogeneous mixture. It is the same throughout, but has two parts—water and salt. Figure 4-5 compares matter and its different divisions.

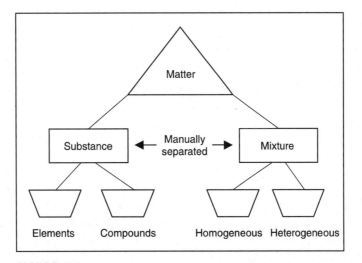

FIGURE 4-5 · Matter can be further broken down into different divisions.

Compounds

Pure chemicals, which can be broken down into simpler chemicals, are known as *compounds*. Commonly, chemical compounds are made up of two elements in specific proportions to each other. Water provides an easy example of a compound. It is composed of the elements hydrogen and oxygen. There are always two hydrogen atoms to one oxygen atom in every water molecule. If the water sample is from the sea or polluted, there may be other chemicals added, but basic water always has the same proportion of hydrogen to oxygen by mass.

What Are Liquids?

Liquids such as water, oil, and alcohol have been known for centuries. They have a measurable volume, but are bendable and can change shape. Unlike rigid solids, they can flow across a tabletop when spilled.

Lava, a cold mountain stream, and mercury are all liquids. They have very different compositions and standard temperatures, but they are all members of an in-between element club, known as liquids. The members of this club are affected by their environment much more that solids. Sometimes just a few degrees of change in temperature can cause an element to slip from the solid club to new membership in the liquid club.

Liquids are often easier to study than solids. They take the shape of their container and flow from one place to another. Great physical force is not needed for liquids to change shape, but they are affected by heat.

Cesium has a particularly great trick of switching from a solid to a liquid. It is solid at room temperature, but when held in the hand and heated to body temperature of 98.6°F, it melts and becomes a liquid.

A chocolate bar does the same thing. At room temperature on a shelf it's a solid, but when held in the hand and heated to body temperature, it melts. Table 4-2 lists some general characteristics of liquids.

Density of a Liquid

Just like solids, density affects the way liquids act in different environments. As we saw earlier in Fig. 4-1, liquids are less tightly packed than solids. The density of a liquid is often compared to water. The density of water is 1.00 g/mL at 4°C. The metric system of measuring liquid density is based on this number. It's easier to figure out whether liquids will mix or not, because two liquids of very different densities don't usually combine.

TABLE 4-2   Liquids are the stars of the in-between club of matter
Liquids
Fixed volume
Loose structure
Cannot be tightly compressed
Have different viscosities or resistances to flow
Held together by surface tension
Miscible or immiscible
Expand and vaporize when heated
Unequal molecular bonding
Medium density

There are exceptions. Very dense ionic solutions like salt water dissolve in fresh water because both are polar. Oil is nonpolar and will not dissolve in water even if the densities are close. Failure to mix is due to different properties, rather than density. In this case, opposites do not attract.

For example, *specific gravity* is often used in brewing beer. The brewer makes a batch of wort (i.e., a sugar broth made from barley extract). With a lot of sugar in solution, the wort is very dense with a high specific gravity. When the brewer adds yeast, the sugar is converted to alcohol. This lowers the wort's density and specific gravity. The brewer keeps testing the specific gravity until it stops changing, which shows the yeast has finished its work. Then, it's time to bottle the beer.

> **Specific gravity** of a material equals the ratio of its density to the density of water at 4°C.

Some physical characteristics important to liquids are *surface tension*, *viscosity*, *vaporization*, *condensation*, *evaporation*, and *boiling point*. These are described below.

Surface Tension

A molecule in the middle of a liquid is attracted equally in all directions by intermolecular forces. However, a strange thing happens at the surface of many liquids. Intermolecular forces holding molecules together are pulled downward by gravity, but there is no equal pull from above where the surface is exposed to air. The unequal downward pull flattens the molecules' shape and allows them to "float" on top of molecules just below them in a container. The floating molecules form a surface film.

When forces between liquid molecules become stronger, so does surface tension. Consider the strong intermolecular forces of water. When dropped as rain through the air, raindrops are roughly spherical (due to surface tension). The forces between molecules pull at one another equally, except those at the drop's surface.

A raindrop is actually not a true sphere. It is stretched slightly longer by gravity. During microgravity experiments in space, water drops form perfect spheres, because gravity doesn't pull them out of shape.

Viscosity

The shape and combination of ions and molecules in solids affect their characteristics. The same is true of liquids. Size, strength, shape, and intermolecular forces of molecules have a big impact on the viscosity of liquids.

? Still Struggling

Think of how ketchup pours slowly, taking its time on the way out of the bottle to your hash browns. Ketchup is thick or *viscous*. Unlike water or orange juice with low viscosities, ketchup has a higher viscosity.

Viscosity is the capability of a liquid to flow or not flow freely at room temperature.

EXAMPLE 4-1

Take a look at the following liquids. Which have high viscosities (hv) and which have low viscosities (lv)?

(1) Pancake syrup
(2) Vinegar
(3) Motor oil

(4) Apple juice

(5) Molasses

(6) Pine sap

☑ **SOLUTION**

Did you get (1) hv, (2) lv, (3) hv, (4) lv, (5) hv, and (6) hv?

Think of the difference between water and honey. Hydrogen and oxygen form permanent covalent bonds, but there are more intermolecular forces of attraction between water molecules called *hydrogen bonding*. Depending on the strength and number of noncovalent interactions between molecules and their surroundings, bonding may be strong or weak, permanent or fleeting. They bind and release often enough to affect the molecule's properties.

Hydrogen bonds are stronger than the forces between organic molecules in proteins. However, huge molecules can have very strong intermolecular forces. When this happens there is an increase in viscosity. The stronger the molecular forces between molecules, the thicker (i.e., more like a solid) a liquid will be. The weaker the molecular forces, the thinner or less viscous the liquid. Table 4-3 compares the different strengths of bonding interactions.

Vaporization

Liquids are greatly affected by the amount of surface area exposed to air. Sometimes, surface molecules reach the energy needed to escape the liquid sample and become a vapor (i.e., gas). The more liquid surface area available, the more likely liquid molecules will escape as a gas. This is called *vaporization*.

TABLE 4-3 The strength of a liquid's bonds has a lot to do with the ability to flow	
	Strength of bonding interactions (kJ/mol)
Ionic bonding	100–1000
Covalent bonding	100–1000
Dipole–dipole intermolecular forces	0.1–10
Dispersion (London) forces	0.1–10
Hydrogen bonding	10–45

Vaporization is the way molecules change from a solid or liquid to a gas.

? **Still Struggling**

In science fiction movies, an alien monster is sometimes "vaporized" by the hero's ray gun. Vaporization uses the same idea. The alien monster's molecules go from a threatening solid form to being scattered harmlessly into the air.

For vaporization to happen, heat is necessary. Heat is energy. Some liquids change to a vapor at room temperature by pulling heat from the environment. When water or perspiration dries (i.e., becomes a vapor) from the skin's surface, it uses body heat. Heat energy from the body gives water molecules the energy to break surface tension attractions and become vapor.

Molar heat of vaporization is the amount of heat it takes to vaporize 1 mol of liquid at a constant temperature and pressure.

Condensation

Similarly, heating water on a hot stove adds vaporization energy. In a covered pot, vaporization only goes on until the space above the liquid is full or so full of liquid vapor there is no more room to expand. The pot's air space is *saturated*.

Through the glass lid of a pot of bubbling soup, for example, you can see water vapor condensing on the lid as droplets. Some of these vaporized molecules fall back into the liquid and are captured through *condensation*.

Condensation, the opposite of vaporization, happens when molecules go from a gas back to a liquid state.

The rate of evaporation and condensation is not steady the whole time in a closed container. It changes constantly. At first, the molecules slowly enter the vapor state. After a while, though, the closed air space becomes packed with molecules and the liquid state absorbs molecules falling back to the surface. When this happens, the liquid sample is said to have reached a state of equilibrium.

While vaporization and condensation rates may differ, the rate of vaporization slows when the rate of condensation speeds up. A liquid is at equilibrium in a closed environment when the condensation and evaporation rates balance.

> When liquid molecules turn to vapor (gas) at equal rates or molecules vaporize at the same rate that they condense, a **dynamic equilibrium** is reached.

Dynamic equilibrium comes about when vaporization and condensation become the same, because forward and reverse processes are taking place at the same time. The exchange is dynamic. The molecules are not stuck, but continue to move back and forth between phases. The overall number of molecules in each phase is constant when the rates are equal. Equilibrium can be remembered in the following way:

> Liquid ⇔ Gas
>
> Vaporization ⇔ Condensation

In the same way, adding heat energizes the molecules and causes them to vaporize much faster. At the same time, condensation speeds up and stability happens at a higher temperature. Figure 4-6 shows rates of vaporization and condensation in a liquid over time.

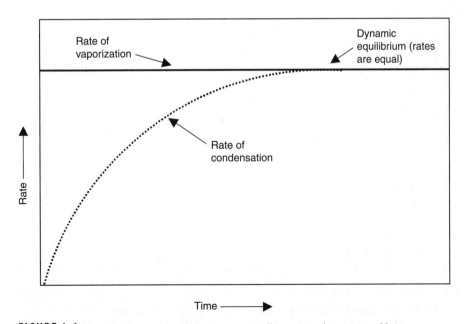

FIGURE 4-6 • Vaporization and condensation eventually come to dynamic equilibrium.

Evaporation

In an open container, vaporization is not limited by the amount of free space above the liquid. It will go on until all the molecules have gone into the air. When this happens, it is called *evaporation*. That is why fish tanks need to have water added every few days, the water molecules escape to the air and the water level goes down. Evaporation is different from vaporization in that it happens below the boiling point of a liquid, at room or outside temperatures.

Lakes and streams get smaller and smaller in hot weather if there is no rain. The water molecules evaporate into the atmosphere and don't come back down again. Water from underground reservoirs or aquifers, are used up and not replenished. In a drought, water rationing is a way to limit water use until the environment changes, condensation happens, and water is returned to the Earth.

Phase Diagram

Chemists use a *phase diagram* when studying the solid, liquid, and gaseous forms of a substance under different pressures and temperatures. It's a graphical tool for illustrating conditions at which a sample's various states are stable. Figure 4-7 shows a phase diagram (not to scale) for water with its triple point shown.

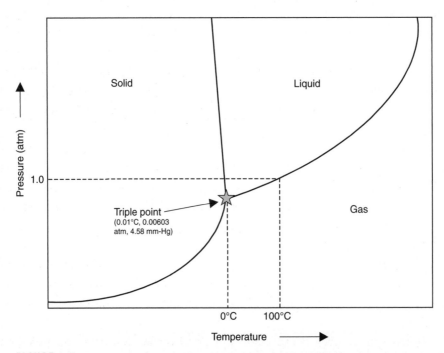

FIGURE 4-7 · The triple point is found at the intersection of the different phases.

The **triple point** is the temperature and pressure point on a phase diagram where three phases of a substance are in equilibrium.

Boiling Point

Temperature has a huge effect on solids and liquids. Since many elements are solid at room temperature (15–25°C), we normally think of them as solids. However, when solids are heated, they melt and become liquids. As we learned earlier, this temperature is the melting point.

Boiling point is to liquids what melting point is to solids. It is the specific point where enough heat energy has been added to the molecules of a liquid sample to allow it to vaporize and become a gas. This heat energy fuels molecules in their journey from a liquid to a gaseous form. They do this when they are at dynamic equilibrium between the two states (i.e., liquid and gas). In other words, at the boiling point, molecules are equally happy as a liquid or gas.

Boiling point is the temperature at which vapor pressure equals atmospheric pressure.

Boiling point ultimately depends on the chemical properties of compounds. Figure 4-7 also shows how increasing temperature provides the heat energy (100°C) for water molecules to break their hydrogen bonding. The water then boils to escape into the air as a vapor. Table 4-4 shows the wide boiling point range for liquids. It makes it easy to see how millions of different liquid processes are affected by temperature.

If a liquid is not pure (e.g., it contains salt), then the boiling point increases. In the case of salt water, salt ions interact with water, so the water molecules need help in the form of heat to become a gas. That is why water is salted when noodles are boiled. The higher temperature helps the noodles cook faster so they don't fall apart in the pot after soaking too long. The same principle applies to other chemicals mixtures.

In the petroleum industry, separation of different parts (i.e., fractions) of crude oil allows the separate collection of many different products. The thick, sticky crude tar is heated in a huge column to separate the different oil components.

As temperature rises, compounds boil off at certain temperatures based on their chemical properties. The heated crude oil products rising up the column

TABLE 4-4 Liquids have a wide range of boiling points depending on structure and bonding

Name	Boiling point (1 atm pressure, °C)
Mercury	356.6
Water	100.0
Ammonia	–33.4
Ethyl alcohol	78.3
Vinegar	118.0
Acetone	56.2
Benzene	80.1
Iodine	184.0
Bleach (chlorine)	–34
Hydrochloric acid	109.0
Xylene	140.0

have different length carbon chains that can be pulled off as pure fractions. This is called *fractionation*. Methane (CH_4), propane (C_3H_8), and butane (C_4H_{10}) oil fractions are purified and collected as fuels.

Gasoline and diesel fuels are larger *hydrocarbon* products (e.g., with 6–12 carbon atoms per molecule) that separate at higher temperatures. Even larger hydrocarbons (e.g., with 20–40 carbon atoms per molecule) are in the form of tar. They are the last to boil away. Generally speaking, the larger the hydrocarbon molecule, then the higher the boiling point. That's how small molecules (e.g., propane) can be separated from larger molecules (e.g., gasoline) within a complex mixture based on boiling points.

Freezing Point

The temperature at which a pure liquid turns into a crystalline solid is called the *freezing point*. This is the temperature when all the energy of the moving liquid molecules is given up and a more ordered (i.e., solid) molecular arrangement is formed. Conversely, a pure solid melts and turns into a liquid at its melting point. Just like the transition from liquid to gas, freezing and melting point depend on a compound's properties.

If a liquid is not pure (e.g., salt water) the freezing point drops. Salt ions make it harder for water molecules to form an ordered structure in the form of ice. More energy must be pulled from the solution, which means a lower temperature is needed for freezing. This is why salt water freezers at a lower temperature than pure water.

When ethylene glycol (CH_2OHCH_2OH) is mixed with water, it serves as an antifreeze in motor vehicles. The combined solution of water and ethylene glycol freezes at a temperature well below water's freezing point (−13°C instead of 0°C). Unlike water alone, a mixture of ethylene glycol and water won't freeze and crack the engine. Now that's cool!

Liquids change states (i.e., freezing and vaporizing) more freely than denser solids. In the next chapter, you will see how gases have even more exciting adventures.

QUIZ

1. **When crude oil is separated into different components of naturally occurring and reactant products it is called**

    A. boiling point.
    B. fractionation.
    C. condensation.
    D. surface tension.

2. **Dynamic equilibrium takes place**

    A. only in solid materials.
    B. as an anaerobic process.
    C. when forward and reverse reactions happen at the same rate and time.
    D. whenever pressure and atmosphere are equal in a semiclosed system.

3. **The temperature at which vapor pressure equals atmospheric pressure is called**

    A. heat of condensation.
    B. boiling point.
    C. crystallization point.
    D. freezing point.

4. **Pure substances are**

    A. homogeneous and have uniform chemical compositions.
    B. found in rare geologically pristine sites.
    C. extremely brittle.
    D. heterogeneous and have unchanging chemical compositions.

5. **Viscosity describes the**

    A. rigid cubic form of a crystalline solid.
    B. rate of heat loss from a sample following its removal from an energy source.
    C. chemical reaction rate of a liquid.
    D. capability of a liquid to flow or not flow freely at room temperature.

6. **Ethylene glycol when mixed with water, is useful as a(n)**

    A. plastic.
    B. detergent.
    C. antifreeze.
    D. dye.

7. **Freezing point is to liquids, as _____ is to solids.**

    A. melting point
    B. vaporization
    C. evaporation
    D. boiling point

8. **The temperature and pressure point on a phase diagram where three phases of a substance are in equilibrium is called the**
   A. acclimation point.
   B. freezing point.
   C. triple point.
   D. composite point.

9. **Specific gravity describes a**
   A. ratio of a sample's density at 20°C divided by the density of water at 4°C.
   B. speed needed to break free from the Earth's gravity and orbit.
   C. sample's volume increase when placed in mercury.
   D. density unit measured in feet per second.

10. **Color, electrical conductivity, and melting point are examples of**
    A. gaseous properties.
    B. chemical properties.
    C. properties of liquids.
    D. physical properties.

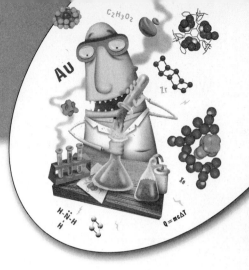

chapter **5**

Gases and the Gas Laws

Gases are a different story from liquids and solids. They have neither form nor specific volume and expand to fill the entire container into which they are placed. At times, they are visible and then disappear. They seem to come from nothing and leave to go nowhere.

CHAPTER OBJECTIVES

In this chapter, you will

- Find out how gases differ from solids and liquids
- Learn the importance of kinetic gas theory
- Discover the impact of pressure and temperature on gases
- Understand the ideal gas law
- Find out how the various gas laws work together
- Become a whiz at Dalton's law of partial pressures

What Are Gases?

Gases have been around and noticed by humans for a long time. Thousands of years ago, ancient Sumerians in the Middle East saw bubbles of gas form when they brewed ales and spiced ciders from fruit and fermented grains. Some ancient peoples believed intestinal gases were somehow connected to the spirit. When the spirit was unhappy, sickness would plague a person with an excess of gas. When pockets of natural gas from the earth were discovered, it was thought the earth's spirit was releasing the people's errors upon the land.

Modern chemists have learned a lot since those days. With the development of precise equipment able to measure tiny samples, accurate information about gases could be gathered and studied. Through years of testing, the *ideal gas law* was developed to explain and predict how gases behave. This chapter will discuss the four parts of this important law.

The gas "club" is a lot more active than the liquid or solid clubs. Gases have livelier characters and no set boundaries. When allowed to escape from a container, they spread out into whatever space is available.

> **Gases** are the least compacted or dense form of matter.

Everyone knows the air we breathe is a gas. Although on very humid summer days, when the air is loaded with water, it feels more like a liquid. Some other common gases include nitrogen, oxygen, carbon dioxide, helium, neon, and argon. Table 5-1 gives some of the general characteristics of gases.

> When gases expand and mix with other gases to fill available space, it is called **diffusion.**

Environmentalists consider diffusion when they measure industrial gas levels in the air. Following a release, they measure gases in parts per million or parts per billion. From these measurements, they are able to calculate release levels and determine whether or not concentrations of diffused pollutants are harmful.

TABLE 5-1   Gases are the wild and care free members of the three matter forms
Gases
No set shape or volume
Expand to fill shape of container
Can be compressed by increasing pressure
Mix completely and spontaneously
Move constantly, quickly, and randomly
Smaller mass gases move more quickly than gases with larger masses
No strong molecular forces between particles
When particles collide, no energy is lost
All collisions are elastic
Low density

Atmosphere

Although the air we breathe may seem limitless, our air only stretches out about 30 km from the surface of the earth. It is only breathable to humans to about 14,000 ft or 4.3 km. The air we breathe, in fact, is not just one gas, but several. Outside and indoor air are made up of roughly 78% nitrogen, 21% oxygen, and 1% argon with a smattering of 3–4% water vapor, carbon dioxide, sulfur dioxide, and others, depending on where you live (Table 5-2).

TABLE 5-2 Many harmless and toxic gases are found in the atmosphere			
Name	**Formula**	**Color**	**Toxicity**
Ammonia	NH_3	Colorless	Toxic
Carbon dioxide	CO_2	Colorless	Nontoxic
Carbon monoxide	CO	Colorless	Very toxic
Chlorine	Cl_2	Light green	Very toxic
Helium	He	Colorless	Nontoxic
Hydrogen	H_2	Colorless	Nontoxic
Hydrogen chloride	HCl	Colorless	Corrosive
Hydrogen sulfide	H_2S	Colorless	Very toxic
Methane	CH_4	Colorless	Nontoxic
Neon	Ne	Colorless	Nontoxic
Nitrogen	N_2	Colorless	Nontoxic
Nitrogen dioxide	NO_2	Reddish brown	Very toxic
Oxygen	O_2	Colorless	Nontoxic
Sulfur dioxide	SO_2	Colorless	Toxic
Sulfur pentafluoride	S_2F_{10}	Colorless	Very toxic

Carbon Dioxide

Scientists have noticed the amount of polluting gases in the atmosphere is rising. For example, back when the pyramids were built, CO_2 levels were around 80–100 ppm. In 1900, CO_2 levels were under 300 ppm. Today's CO_2 levels are nearing 400 ppm, which is the highest they've been over the past 650,000 years. That's more than 4 times what it used to be.

Since the human body doesn't do well breathing low levels of oxygen, this is a huge problem. While air is no longer pure in many parts of the world, rising carbon dioxide levels add to the problem of global warming. CO_2 is the most abundant greenhouse gas. We will look at this more closely in Chap. 14.

Kinetic Energy and Gas Theory

Gas molecules are constantly on the move. They are always bouncing off each other and anything else they come into contact with. They are super charged with energy. When scientists talk about this crazy gas motion, they call it *kinetic energy*.

The kinetic energy of gases can be calculated as equal to one-half the mass (m) of a sample multiplied by the velocity squared (v^2) (see below). If a scientist has a contained sample of a known mass with its molecules bouncing all over the place, its kinetic energy can be calculated using this equation. In fact, knowing a gas' energy helps scientists predict how it will behave.

Chemists calculate the kinetic energy of gases based on their temperature, which in turn affects their velocity (i.e., speed). The average kinetic energy of a gas molecule depends on the absolute temperature of the gas.

$$\text{Kinetic energy} = \tfrac{1}{2}\,mv^2$$

Graham's Law

Thomas Graham was a Scottish physical chemist interested in the movement or kinetics of gases. When a German chemist, Johann Döbereiner reported that hydrogen gas slipped through a crack in a bottle faster than the air replacing it, Graham decided to see for himself. To check it out, Graham measured the rate of gases leaking through a number of differently sized "holes." In this way, he slowed the process enough to take measurements and explain his observations. Graham's experiments showed the gas effusion (i.e., leaking) rate is inversely proportional to the square root of the mass of its particles. The equation for Graham's law is:

$$r_1/r_2 = \sqrt{M_2/M_1}$$

where, r_1 = rate of effusion of the first gas (volume or number of moles/unit time)
   r_2 = rate of effusion for the second gas
   M_1 = molar mass of the first gas
   M_2 = molar mass of the second gas

(Note: See Chap. 6 for more information on molecular weight.)

Atmospheric Pressure

As we've learned, the Earth's atmosphere is a mixture of different gases. Gravity affects the force by which gas molecules constantly collide with everything. Its pull on gas molecules decreases as molecules get farther and farther away from the planet. Similarly, gas molecules' weight also drops without gravity's constant drag. It's the same reason why astronauts are weightless in space and weigh only about a third of their regular (Earth) weight on the moon.

> **Atmospheric pressure** (atm) equals the weight of air per unit area.

Barometer

Pressure is measured two different ways depending on whether we are talking about atmospheric pressure or the pressure within a closed container (e.g., a gas cylinder). For atmospheric pressure, a *barometer* is used. This involves a straight, hollow tube, sealed at one end while the other end is immersed in mercury. Gravity forces the mercury level inside the tube down while atmosphere gases at mercury's surface force the mercury level up the tube. Figure 5-1 illustrates how a barometer is constructed. When conflicting forces reach a balance in the tube, mercury's height indicates the atmospheric pressure.

Manometer

To measure the pressure inside a container, a *manometer* is used. The concept is the same as with a barometer, but the system is set up differently. Figure 5-2 illustrates a manometer.

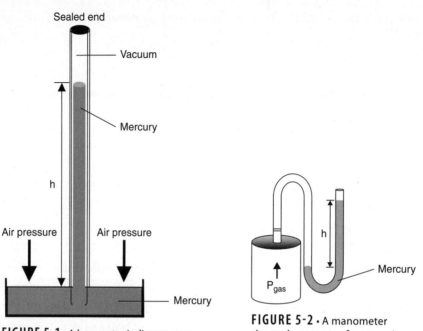

FIGURE 5-1 · A barometer indicates atmospheric pressure.

FIGURE 5-2 · A manometer shows the amount of pressure in a container.

Because the first experiments to find differences in pressure were performed with mercury and tall glass tubes by Evangelista Torricelli, an Italian scientist, the SI standard unit for pressure was called the *torr*.

Experimenters found the experimental column height was 760 mm high and equal to 1 atm at sea level. However, atmospheric pressure changes depend on how low (e.g., below sea level) or how high (e.g., mountain peak) you are. By moving around, pressure moves up or down, respectively. Weather, such as tornadoes and hurricanes, also changes atmospheric pressure.

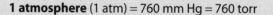

1 atmosphere (1 atm) = 760 mm Hg = 760 torr

Gravity's force on a column of air applies a pressure at the earth's surface. This pressure changes according to elevation and weather, but is commonly written as 760 mm Hg or 1 atm or 101 kPa. [Note: When engineers and mechanics calculate pressure, they often use the units, pounds per square inch (psi).]

Empirical Gas Laws

Boyle's Law

In 1662, Robert Boyle, an Irish chemist, wanted to find out how gases were affected by outside factors. First, he bent a glass tube into a hook shape and sealed it one end. Then, he poured mercury into the tube. Boyle discovered that a small volume of air became trapped at the tube's sealed end.

Because of mercury's density, Boyle realized the more mercury he poured into the tube, the harder it pushed against the trapped air at the end of the tube. After adding enough mercury to compact the trapped air into half of its original space, Boyle realized the more he increased the pressure against the trapped gas volume, the more it was compressed when the temperature was constant.

When Boyle doubled the pressure, the gas volume was reduced by half. When the pressure was tripled, the gas volume was condensed into one-third of its original volume. This inverse relationship became Boyle's law and eventually the first part of the ideal gas law.

> In **Boyle's law**, when temperature is held constant, a volume of gas is inversely proportional to the pressure; $V \propto 1/P$

This relationship can be written in an equation. For an experiment, a gas exists in an initial state (i) and changes to a final state (f). Based on Boyle's law, volume changes affect pressure and vice versa. As initial conditions become final conditions, the difference between pressure (P) and volume (V) at constant temperature can be written into a useful equation and rearranged to solve for an unknown variable:

$$P_i V_i = P_f V_f$$

EXAMPLE 5-1

Find the amount of oxygen in a container with an initial volume of 4.0 L. The initial pressure of the contained gas is 1470 psi when at 25°C. If the pressure is changed to 1 atm, what is the oxygen's final volume if the temperature doesn't change?

 SOLUTION

$$P_i = 1470 \text{ psi and } P_f = 14.7 \text{ psi}$$
$$V_i = 4.0 \text{ L and } V_f = [x] \text{ L}$$
$$4.0 \text{ L} \times 1470 \text{ psi} = V_f \times 14.7 \text{ psi}$$
$$V_f = 4.0 \text{ L} \times 1470 \text{ psi}/14.7 \text{ psi} = 400.0 \text{ L}$$

(Note: *x* is unknown liters)

This is why many industrial gases are stored at high pressure, because more gas molecules can be packed into a smaller volume.

Charles's Law

The second part of the ideal gas law explains the effect of changing temperatures on a gas. Jacques Charles, a French physicist, carried out gas studies in 1787. When he wasn't in his lab, Charles was out hot air ballooning. He was known as one of the best balloonists in France. In fact, Charles was the first to use helium to inflate a balloon capable of carrying passengers.

If we think of what happens with gas atoms, it's easy to remember Charles's law. Temperature energizes a sample. When atoms are heated, they move and collide faster. This effect was explained by kinetic gas theory.

Remember, kinetic energy (KE = ½ mv^2) increases, but the mass stays the same, so the velocity has to increase. Like a happy puppy that can't stay in one place, heated atoms get crazy wild and hit the sides of their containers harder and more frequently causing it to expand.

> In **Charles's law**, when pressure is held constant, a volume of gas is directly proportional to the kelvin temperature; $V \propto T$

Charles used his interest in science and ballooning to test ideas about the relationship between a gas and temperature. He found that the more a gas was heated, the more its volume increased. Like Boyle, he was able to write out a practical equation describing the change in volume (V) and temperature (T) at constant pressure from an initial state (i) to a final state (f).

$$V_i/T_i = V_f/T_f$$

EXAMPLE 5-2

If a 1-L balloon is flexible and can expand with increased temperature (from 20°C to 45°C), find the volume of the balloon after it is heated. (Tip: Add 273 to the Celsius temperature to get everything into kelvin.)

SOLUTION

$$V_i = 1.0 \text{ L}, T_i = 27°C, V_f = ? \text{ L}, T_f = 1500°C$$

Remember to add 273 to get the temperature in kelvin.

$$T_i = 20 + 273 = 293 \text{ K}, T_f = 45 + 273 = 318 \text{ K}$$
$$1.0 \text{ L}/293 \text{ K} = V_f/318 \text{ K}$$
$$V_f = (1.0 \text{ L}/293 \text{ K}) \times 318 \text{ K}$$
$$V_f = 1.09 \text{ L}$$

Gay-Lussac's Law

Around the same time Charles was ballooning and experimenting in France, another French scientist, Joseph Gay-Lussac began studying the connection between temperature and gas pressure. His research added the third part of the ideal gas law. Gay-Lussac discovered that as temperature and kinetic energy increases, pressure increases as well.

In **Gay-Lussac's law**, when volume is held constant, gas pressure is directly proportional to the kelvin temperature in: $P \propto T$.

Again, gas atoms go wild in a constant volume. As the temperature increases, the pressure increases and the atoms collide with the container walls faster and harder. This raises the kinetic energy. The following equation describes what happens when pressure (P) and temperature (T) change at constant volume.

$$P_i/T_i = P_f/T_f$$
$$P_f = P_i \times T_f/T_i \underline{\textbf{or }} T_f = T_i \times P_f/P_i$$

? Still Struggling

When you read the label of a pressurized spray paint can, you will probably see a warning to keep the can from coming in contact with heat. You can thank Gay-Lussac for this warning label. When a can is heated enough, the pressure increases and the can will explode. Besides being very dangerous, you will paint everything in sight. The take-home chemistry message is, NEVER HEAT A SPRAY CAN!

EXAMPLE 5-3

A pressurized (875 torr) room temperature (25°C) hair spray can [i.e., 15 ounces (oz)] is heated to 1500°C in a house fire. Find the pressure inside the can before it explodes.

SOLUTION

$$P_i = 875 \text{ torr}, T_i = 27°C, P_f = ? \text{ torr}, T_f = 1500°C$$

Remember to add 273 to get the temperature in kelvin.

$$T_i = 27 + 273 = 300 \text{ K}, T_f = 1500 + 273 = 1773 \text{ K}$$
$$875 \text{ torr} \times 1773 \text{ K} = P_f \times 300 \text{ K}$$
$$P_f = (875 \text{ torr} \times 1773 \text{ K})/300 \text{ K}$$
$$P_f = 5171 \text{ torr}$$

The pressure at this high heat is nearly 6 times what the spray can is designed to hold!

Combined Gas Law

The last three gas laws have been integrated into the combined gas law. This equation, makes it possible to figure out nearly every temperature, volume, and pressure of a sample if you know the other constants. The combined gas law is made up of the general rules for temperature, volume, and pressure described in Boyle's, Charles's and Gay-Lussac's laws to describe what happens when conditions for a gas change.

It is written as follows:

$$P_i V_i/T_i = P_f V_f/T_f$$

When you know five of the six values, you can figure out the missing value. If any of the variables (pressure, temperature, and volume) do not change, then the equation simplifies to one of the three previously described gas laws.

◻ EXAMPLE **5-4**

If a bicycle tire has a volume of 0.5 m³ at 20°C and 1 atm (760 torr), figure out the volume when taken into the mountains where the pressure is 720 torr and temperature is 14°C.

✔ SOLUTION

$V_i = 0.5$ m³, $T_i = 20°C$, $P_i = 760$ torr, $P_f = 720$ torr, $T_f = 14°C$, $V_f = ?$ m³

$T_i = 20 + 273 = 293$ K, $T_f = 14 + 273 = 287$ K

$$P_i V_i/T_i = P_f V_f/T_f$$

0.5 m³ × 293 K × 760 torr = V_f × 287 K × 720 torr

$V_f = (0.5$ m³ × 293 K × 760 torr$)/(287$ K × 720 torr$)$

$V_f = 0.54$ m³

As the temperature and pressure decrease, the tire volume is less compressed and feels mushy. To get the same smooth ride from the tires, pump them up a bit.

Avogadro's Law

Before we can put together the ideal gas law, there is a fourth part to be reviewed. In 1811, Amedeo Avogadro, an Italian physicist, noted that when equal volumes of gases at the same temperatures and pressures are present, the gases have equal numbers of molecules.

Just like the previous gas laws, Avogadro's law can be expressed as an equation to understand the relationship between volume (V) and the number of molecules (n) when temperature and pressure remain unchanged.

$$V_i/n_i = V_f/n_f$$

> **Avogadro's number** (L) is a constant number of atoms, ions, or molecules in a sample. It is equal to the number of atoms in 12 g of carbon-12 or 6.022×10^{23}.

To give you an idea of the unthinkable number of molecules contained in Avogadro's number, collect a whole lot of hazelnuts (each roughly 2 cm in diameter) and cover the area of the United States. To equal Avogadro's number, the hazelnut layer would have to be over 100 km (i.e., 70 miles) deep. The importance of this number and a unit called the *mole* will be discussed in Chap. 6.

Ideal Gas Law

An ideal gas meets all the gas laws' rules. When all four gas laws (Boyle's, Charles's, Gay-Lusac's, and Avogadro's laws) are combined, we get the *ideal gas law*. These laws were discovered by different scientists at different times, but all add up to explain the strange and amazing things gases do in different conditions. To make the law useful, a constant (R) was included in the equation that considers all four gas laws.

$$PV = nRT$$

where, P = gas pressure (i.e., atmosphere, torr, mm Hg, Pa, etc.)
V = volume (i.e., liters, milliliters, etc.)
n = number of molecules/moles of gas at constant pressure and temperature
R = ideal gas constant, 0.0821 L·atm/K·mol
T = kelvin temperature

Standard Temperature and Pressure

When studying the gas laws, you may encounter standard temperature and pressure or *STP*. Table 5-3 lists the values for temperature and pressure that are called *gas standards*. Scientists use these conditions to simplify experiments with less variables.

The activity of gases can be calculated using a lot of different ratios. If you know the temperature, pressure, and volume of a gas, then lots of different ideas can be tested. The gas club is definitely the most changeable of the solid, liquid, and gas forms of matter.

TABLE 5-3 Using standard temperature and pressure units for gases makes conversions simpler	
Standard temperature and pressure	**(STP)**
1 standard temperature	0°C
1 standard temperature	273 K
1 standard pressure	1 atm
1 standard pressure	760 torr
1 standard pressure	14.7 psi

Molar Gas Density and Volume

Gas molecules prefer to be as far as possible from each other without wanting either a special shape or volume. To study gases, it's helpful to know gas density. The density of a gas is measured in mass per unit volume ($D = m/V$) with units like g/L. To figure out the molar volume of a gas, multiply gas density times its molar mass. Molar mass is discussed more fully in Chap. 6.

$$\text{Molar volume } (V_m) = D \times m$$

For example, the molar volume of oxygen can be calculated from the density of oxygen at STP (1.43 g/L) and its molar mass (32.0 g/mol). To figure out the molar volume of oxygen at STP, use the following formula.

$$32.0 \text{ g/mol} \times 1 \text{ L}/1.43 \text{ grams} = 22.4 \text{ L/mol}$$

Because Avogadro's number is so huge, every gas works out to be very close to 22.4 L/mol. In other words, all gases have basically the same molar volume at STP or 22.4 L/mol.

Van de Waals' Equation

In the ideal gas law, gas molecules are thought to be so small and without any reactions between molecules that size doesn't matter. However, real gases aren't always ideal. As discussed for Graham's law, size matters when studying gas properties.

Additionally, when considering low temperature or high pressure, gases behave less ideally. At high pressure, gas molecules are packed very tightly and can stick together. When this happens, any bouncing around from kinetic energy is slowed.

In 1873, Johannes Diderik van der Waals modified the ideal gas law to correct for these nonideal interactions in the *van der Waals equation*.

$$(P + n^2a/V^2)(V - nb) = nRT$$

where P = gas pressure (atm)

V = gas volume (L)

n = # of moles of gas (mol)

T = absolute gas temperature (K)

R = gas constant, 0.0821 L-atm/mole-K

a = a constant, specific for each gas, which accounts for attractive forces between molecules

b = a constant, specific for each gas, which accounts for the volume of each molecule

Dalton's Law of Partial Pressures

John Dalton had an idea about how gas pressure works. Like Gay-Lussac, Dalton had a hobby. He was interested in meteorology—the study of weather. While studying weather changes, Dalton did some experiments with vapor pressure. He found that, like people, gases are unique and behave in their own way.

Because of this uniqueness property, he noticed each gas compressed at its own pressure even in a mixture. When Dalton tested three different gases at a constant temperature, he found the total pressure of the three gases was equal to the sum of the mixed individual gases. This general rule became known as Dalton's law of partial pressures.

> **Dalton's law of partial pressures** states when a gas is mixed with one or more different gases, the pressure of each gas adds together to get the mixture's total pressure.

Dalton's law is probably the easiest of all the gas laws to remember. It is written below:

$$P_{total} = P_1 + P_2 + P_3 + P_4 + P_5 + \cdots$$

EXAMPLE 5-5

If the atmosphere of mixed gases on an alien planet is made up of oxygen ($p = 0.3$ atm), argon ($p = 0.1$ atm), nitrogen ($p = 0.8$), and neon ($p = 0.01$), what is the total pressure of the atmospheric gases?

SOLUTION

$$P_{total} = P_1 (0.3) + P_2 (0.1) + P_3 (0.8) + P_4 (0.01)$$
$$P_{total} = 1.21 \text{ atm}$$

Following Dalton's law, the sum of the different pressures in the alien atmosphere is greater than the pressure of any single gas alone.

Vapor Pressure

As discussed in Chap. 4, liquids vaporize into the gas phase, which adds another partial pressure to Dalton's equation. For example, on a humid day the air is saturated with water vapor. This partial pressure from vaporized water is called the *vapor pressure*. This property is not limited to water. It applies to any liquid with molecules escaping to the gas phase. Because vapor pressure involves the transition between liquid and gas, it is affected by temperature. As temperature rises, vapor pressure increases. This makes sense, because there are more molecules in the air as a gas than as a liquid.

As you have seen, gases have distinct properties quite different from solids and liquids. It took many years and several scientists of different backgrounds and interests to pull all the factors affecting gases together into useful laws and equations.

QUIZ

1. Kinetic energy can be found with which of the following equations?
   A. Kinetic energy $= mc^2$.
   B. Kinetic energy $= mv^2$.
   C. Kinetic energy $= P/v$.
   D. Kinetic energy $= \frac{1}{2}\, mv^2$.

2. Boyle's law explains
   A. the relationship between metals and nonmetals.
   B. that when temperature is held constant, a volume of gas is inversely propor-
      tional to the pressure.
   C. the relationship of atmospheric pressure and malleability.
   D. the equation $V_1/T_1 = V_2/T_2$.

3. Which law explains that when volume is held constant, a gas' pressure is directly
   proportional to the kelvin temperature?
   A. Gay-Lussac's law
   B. Boyle's law
   C. Cole's law
   D. Charles's law

4. Dalton's law of partial pressures states that when
   A. gases are combined, only ideal gases will combine at any one time.
   B. a gas volume expands, pressure is not a factor.
   C. a gas mixes with one or more different gases, the pressure of each gas will
      combine to get the total pressure of the mixture.
   D. more than one gas mixes with one or more different gases, the total pressure
      of the mixture will be the same as the heaviest gas.

5. Diffusion happens when
   A. two liquids meet.
   B. the barometric pressure is high.
   C. a gas is compressed in a column with mercury.
   D. gases expand and mix with other gases to fill available space.

6. In the equation, $PV = nRT$, P is equal to
   A. pressure.
   B. volume.
   C. number of moles of gas.
   D. temperature.

7. The law of partial pressures is written with which of the following equations?
   A. $P_{total} = (P_1 + P_2)/P_3$
   B. $P_{total} = P_1 + P_2 + P_3$

C. $P_{total} = (P_1 + P_2) \times \mu$

D. $P_{total} = 2 \times (P_1 \times P_2 \times P_3)$

8. **Charles's law explains**

   A. that gas molecules are seldom in motion.

   B. that gases at the same temperature and pressure have decreasing volume.

   C. that the more a gas is heated, the more its volume increases at constant pressure.

   D. the relationship between gas pressure and volume.

9. **Avogadro's number is a constant number of atoms, ions, or molecules in a sample and equals**

   A. 5.250×10^{10}.

   B. 6.100×10^{18}.

   C. 6.022×10^{23}.

   D. 6.122×10^{22}.

10. **Molar volume describes the density of a gas multiplied by its**

    A. pressure.

    B. temperature.

    C. atomic number.

    D. molar mass.

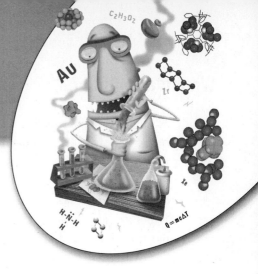

chapter **6**

Solutions

Just like soup on a restaurant menu, solutions are a common part of any lab. Scientists can more easily control the amount of chemicals added to reactions, since solutions provide a good place for reactions to occur. This chapter deals with the specifics of concentrated chemical solutions in the lab and the environment.

CHAPTER OBJECTIVES

In this chapter, you will

- Learn how chemists make solutions
- Find out why some solutions mix and others don't
- Learn how to change the concentration of solutions
- Find out how to convert mass to moles
- Understand parts per billion in the environment

What Is a Solution?

A *solution* is a homogeneous mixture of a *solute* (i.e., element or compound being dissolved) in a *solvent* (i.e., solution dissolving the solute molecules). Commonly, the larger sample is the solvent and the smaller sample is the solute. When salt is dissolved in water, the salt is the solute and water, sometimes called the *universal solvent,* is the solvent.

Solutes and solvents can be elements or compounds. For example, painters use turpentine to thin oil–based paints to clean paint from brushes. Turpentine serves as the solvent, while the paint is the solute.

> A **solute** is an element or compound dissolved to form a solution. A **solvent** is a liquid into which an element or compound is added to form a solution.

To understand how atoms, elements, and molecules interact in solutions, it is important to understand *concentration.* Concentration equals the number of solute molecules in a solvent. The amount of solute molecules determines the solution's function. Consider salt water; if a hospital patient needs more fluid in the blood, a physician may prescribe a saline (diluted salt water) solution of 0.85% NaCl to be given intravenously. By contrast, if a fisherman wants to preserve codfish, he places it in salt brine, which is a 30–50% NaCl solution. These solutions have different properties and functions. Confusing them would kill the patient or lead to decaying fish.

Solubility Rules

Solubility is the ability of a compound to dissolve into another. When binding and becoming more stable, compounds separate into individual ions. Then, at some point, a compound stops dissolving in a solvent.

> Solubility of a solute is written as the ratio of grams of solute per 100 g of water (g solute/g solvent) at a set temperature.

Like dissolves like. Polar liquids, like water, are able to dissolve polar and ionic solutes. Nonpolar liquids, like oil and paint thinner, are able to dissolve nonpolar solutes. We will learn more about polar and nonpolar bonds in Chap. 8.

As discussed in Chap. 4, a solution becomes saturated at the point when no more of a substance can be dissolved into it without a solid precipitate forming.

Miscibility

When two or more liquids readily form a solution, they are called *miscible*. Isopropyl alcohol is miscible in water and when the two are mixed, it is known as rubbing alcohol. There is even an entire subject in physics called *fluid dynamics* that tests the miscibility of fluids and the way they flow at different concentrations.

When two liquids don't mix, like oil and water, they are *immiscible*. Picture a lava lamp. Large round globules of "lava," made of a colorful, compounded wax, float lazily around in a specially formulated liquid. This mixture, heated by a light bulb at the lamp's base, causes the lava to heat and expand until it becomes less dense than the surrounding liquid. When this happens, the lava rises to the top of the lamp where it slowly begins to cool. After the lava cools sufficiently, it sinks to the bottom, is reheated and the process repeats.

Colloids

A *colloid* is like a solution in some ways, but not others. A colloid is a mixture of large particles of one compound blended into another compound. The large particles are not individual molecules like those of a solute in a solution. Though similar, colloids are mixtures of compounds (e.g., gases, liquids, or solids) not actually solutions.

?

Still Struggling

There are two parts to a colloidal mixture, the *dispersed* and *continuous phases*. Chocolate chip cookie dough is like a colloid. The continuous phase (e.g., dispersing medium) is the substance in a colloidal mixture in the greater amount, like the cookie dough. The dispersed phase is the substance in the colloidal mixture in the smaller amount, like the chocolate chips.

One way to tell the difference between a colloid and a solution is what is known as the *Tyndall effect*. John Tyndall, a British scientist in the 1800s, discovered that light is scattered by small particles in suspension, such as a colloid. This process doesn't happen with a true solution, because the solute molecules are too small.

There are different types of colloids. An *aerosol* like smoke is made up of particles dispersed through a gas. Similarly, an *emulsion* is composed of liquid droplets dispersed through another liquid like mayonnaise (i.e., oil dispersed in water).

What Is a Mole?

An important term in chemistry and in making solutions is the *mole*. To a chemist, a mole (mol) is not a small, furry animal that lives in dark, underground burrows, but an International Standard (SI) unit referring to the amount of a chemical present in a sample. As discussed in Chap. 5, Avogadro developed a standard for measuring the number of molecules, the mole. Avogadro's number for a mole of sample is equal to 6.02×10^{23} atoms or molecules. This unit is important for solutions.

A mole describes a particular number of things, like a dozen eggs equals 12 eggs, a bushel of oats equals 32 pounds, and a mole of propanol equals 6.02×10^{23} propanol molecules.

> A **mole** (mol) equals the same amount of molecules as the number of atoms in exactly 12 g of carbon-12 (i.e., the amount of sample containing Avogadro's number of molecules).

Calculating a Mole of Sample

A high-speed computer can count all the fish in the sea in less than a second, but it would take over a million years to count a mole of fish. Finding the number of molecules in a sample is a similar problem. Scientists use Avogadro's number to determine these amounts.

EXAMPLE 6-1

If you wanted to find the number of molecules in 1.35 moles, you would multiply the number of moles by Avogadro's number.

$$6.022 \times 10^{23} \times 1.35 \text{ moles} = 8.1297 \times 10^{23} \text{ molecules}$$

Molar Mass

The *molar mass* (MM) of an element is the mass, in grams, equal to the atomic and formula masses of those elements and compounds. It is measured in grams/mole. Molar mass, also known as an element's *atomic weight* is measured out in grams equal to one mole of the element's atoms. On the Periodic Table, molar mass is the average of the mass numbers of all the known isotopes for a specific element weighted by their percent of abundance.

Molecular weight (MW) equals the sum of the atomic weights of atoms in a molecule of a substance (e.g., atomic mass units or amu).

Still Struggling

The molecular weight of sodium hydroxide (NaOH) is $(1 \times 22.99$ amu for the Na atom$) + (1 \times 16.0$ amu for the O atom$) + (1 \times 1.0$ amu for the H atom$) = 39.99$ amu.

Conversion of Mass to Moles

In the course of an experiment, sometimes a researcher needs to figure out an element's mass. For example, he/she might be asked to calculate the mass of 2 moles of potassium (K). To do this, a conversion factor to change moles to mass is needed. The conversion of moles to mass is grams/mole.

$$2.0 \text{ mol·K} \times 39.0 \text{ g/mol·K} = 78.0 \text{ g}$$

How about finding the number of moles in 124 g of calcium? Mass to moles then would be moles/gram.

$$124 \text{ g} \times 1 \text{ mol Ca/40.1 g} = 3.09 \text{ mol Ca}$$

Using these simple conversion factors, you can find a compound's *molar mass*. The following solution of $CaCO_3$ shows how it works:

$$\text{Ca 1 mol Ca} \times 40.1 \text{ g/mol Ca} = 40.1 \text{ g}$$
$$\text{C 1 mol C} \times 12.0 \text{ g/mol C} = 12 \text{ g}$$
$$\text{O 3 mol O} \times 16 \text{ g/mol O} = 48.0 \text{ g}$$
$$\text{Molar mass} = 100.1 \text{ g/mol CaCO}_3$$

To see the percentage of each element in $CaCO_3$, take the mass of each element in a 1-mol sample and divide by the molar mass.

$$\% \text{ Ca 40.1 g Ca/100.1 g CaCO}_3 \times 100 = 40.1\%$$
$$\% \text{ C 12 g C/100.1 g CaCO}_3 \times 100 = 12.0\%$$
$$\% \text{ O 48 g O/100.1 g CaCO}_3 \times 100 = 48.0\%$$

Molarity

What happens when you put a mole in a liter of water? You get a 1 molar solution! Moles per liter, or *molarity*, is a helpful way to describe chemical solutions. A solution's units provide information about the amount of chemical molecules in the solution and its strength. More solution, more molecules. Simple.

> **Molarity** (M) equals concentration. M equals the number of moles of solute (*n*) per volume in liters (*V*) solution.

Molarity is the most common way scientists describe solutions.

$$\text{Molarity} = \text{grams}/(\text{molecular weight} \times \text{volume})$$
$$\text{Moles/liter} = \text{grams}/[(\text{grams/mole}) \times \text{liters}]$$

Percent Solution

Relative amounts of solutes and solvents are also important. If the volumes of two liquids are known (e.g., water and isopropanol), then the solution can be described as *volume-by-volume percent* (v/v%) or simply *volume percent*. The volume of a solute in a 100-mL solution is percent volume.

Rubbing alcohol is usually 70% (v/v), which means 70 mL of isopropanol was added to 30 mL of water to get the final 100 mL solution. If the solute is a solid, then a *mass-by-volume percent* (m/v%) is used to describe the solution. The mass-by-volume percentage is the amount of solute in grams added to 100 mL of solvent.

The brine solution (30% NaCl) used by the fisherman mentioned earlier was made by adding 30 g of NaCl to 100 mL water. If relative masses are considered, the concentration of the compounds could be expressed as *mass-by-mass percent* (m/m%) or simply mass percent. To do this calculation, solute mass is divided by the mass of the solution (i.e., solute plus solvent) and multiplied by 100 to get a percent as in Example 6-2. For all percent solutions, the notation, v/v, m/v, or m/m, is always included to avoid confusion about what the percentage actually means for that solution.

EXAMPLE 6-2

What is the weight percent of a sucrose solution if 4.7 g of sucrose is dissolved in 145.2 g of water?

SOLUTION

$$\% \text{ by mass (mixing in solute)} = \frac{\text{mass of solute mixed}}{\text{mass of total solution}} \times 100\%$$

First, what is the total mass of the solution? Well, 4.7 g of sucrose (solute) are combined with 145.2 g of water (solvent) to make a 149.9-g solution. Next, divide the mass of solute by the mass of the solution and multiple by 100%, or 4.7 divided by 149.9 times 100 to get a 3.1% sucrose solution (m/m%).

Parts Per Million and Beyond

Another way scientists talk about concentrations is a *parts per* notation. Parts per is an easy notation used for low and extremely low concentrations. It's a way of talking about relative proportions of chemicals like the mass-by-mass percentages discussed in the previous section. For example, 1% w/w means 1 g of a compound per 100 g of sample, which may also be written 1 part per hundred (1 pph). Usually concentrations are much lower than pph, so that *parts per million* (ppm) and *parts per billion* (ppb) are commonly used. Common abbreviations for the ratio of chemicals are shown below:

ppm—parts per million (10^6)

ppb—parts per billion (10^9)

ppt—parts per trillion (10^{12})

ppq—parts per quadrillion (10^{15})

"Parts per" notation is useful in environmental studies to discuss the presence of chemicals around us. In congested and metropolitan areas, industries and vehicles release atmospheric pollutants such as soot particles and sulfur dioxide (SO_2). These air pollutants along with nitrogen monoxide (NO) and dioxide (NO_2) are major concerns for human health impacts. More about this will be discussed in Chap. 14.

Small pollutant levels are cleansed by Mother Nature, but when there is too much pollution, there is big trouble. Atmospheric pollutants merge with water in the air to form acids that eventually rain down on everyone and everything. Acid rain acidifies our drinking water, kills animal and plant life including crops, and corrodes buildings, statutes, cars, and a lot of other things.

Small particles associated with atmospheric gases also contribute to the formation of acid rain as well as making breathing problems (e.g., asthma) worse.

However, it is important to know that even though industrial release is measured in ppm or ppb, it may or may not be toxic. A lot depends on the specific structure and binding properties of the released chemical or compound,

For example, if you swallow 300 ppb of inorganic arsenic in a glass of polluted water, you may get a stomachache, nausea, vomiting, and diarrhea along with possible heart problems. However, a much higher arsenic dose (e.g., 70–200 mg or 1 mg/kg/day) is lethal in adults.

In the case of hydrogen cyanide, an exposure of only 200–500 ppm can cause loss of consciousness after only a couple of breaths. So it is critically important to know the *parts per* concentration of a toxic release as well as the pollutant's characteristics.

Try a *parts per* calculation in the following example.

EXAMPLE 6-3

Parts per million can be found by multiplying the ratio of the mass of solute to mass of solution by 10^6 ppm instead of 100%.

$$\frac{4.2 \times 10^{-3}\text{g (solite)}}{1.0 \times 10^3\text{g (solution)}} \times 100\% = 4.2 \times 10^{-4}\%$$

$$\frac{4.2 \times 10^{-3}\text{g}}{1.0 \times 10^3\text{g}} \times 10^6 \text{ppm} = 4.2 \text{ ppm}$$

$$\frac{4.2 \times 10^{-3}\text{g}}{1.0 \times 10^3\text{g}} \times 10^9 \text{ppb} = 4.2 \times 10^3 \text{ppm}$$

In the above example, the chemical release, would probably be reported in ppm.

Table 6-1 lists some ppm levels of common chemical contaminants.

TABLE 6-1 Pollutants are most often measured in parts per million in the air and water	
Pollutant	**Toxic levels (ppm)**
Arsenic in playground soil	10.0
Arsenic in mine tailings (toxic)	1320
Diethyl ether	400
Trihalomethane (in water)	0.10
Nitrate (in water)	10.0
Nitrite (in water)	1.0
Mercury (in water)	0.002
Cadmium (in water)	0.005
Silver (in water)	0.05

Changing the Concentration

A solution's concentration changes when either the number of solute or solvent molecules changes. If the amount of solute increases or solvent decreases, then a solution becomes more concentrated.

If solute ⬆ *or* solvent ⬇

Then, solution = ⬆ concentration

For example, the Dead Sea is the lowest body of water on the Earth and nothing drains out of it. Everything flowing to the Dead Sea is stuck there, except the water. Because of high regional temperatures 86°F (30°C) during winter and 104°F (40°C) during summer, water evaporates quickly. Water leaves, but the mineral salts remain. Over a very long time, the mineral salt concentrations have become extremely high.

The addition of a solvent, such as water, dilutes a solution and lowers its concentration. This principle is commonly used in the lab. A scientist often makes concentrated solutions and then dilutes them as needed. Not only does the scientist save storage space, s/he can make a variety of solutions without having to weigh out lots of chemicals.

Figuring out how to dilute a solution takes some math, but it's an important skill in science. In a dilution experiment, concentration and volume change, but the solute amount does not. Concentration is the amount of chemical per unit volume, so the amount of chemical is equal to the volume times the concentration. This is shown in the equation below for a concentrated (c) starting solution. The equation for the final dilute (d) solution is also shown.

n = the amount of chemical, V = volume, C = concentration

$$C_c \times V_c = n$$
$$C_d \times V_d = n$$

Because both solutions have the same chemical amount (n), the equations can be solved for the concentration of the dilute solution.

$$C_c \times V_c = C_d \times V_d$$
$$C_d = C_c \times V_c / V_d$$

Every chemist must know how to make a less concentrated solution from a concentrated one. They learn from practice.

EXAMPLE 6-4

If an experiment calls for 1 M of hydrochloric acid (HCl) and all you have in the lab is 12 M HCl, could you still do the experiment? Could you use what you had on hand?

SOLUTION

Yes. Just prepare the 1 M HCl by measuring a 1/12 volume or 82 milliliters (mL) of the concentrated solution into 1 L of distilled water. The final concentration is equal to 1 M HCl.

EXAMPLE 6-5

How about nitric acid (HNO_3)? What if you needed 1 M of nitric acid for an experiment and only had concentrated nitric acid (16 M) on hand?

SOLUTION

Measure out 63 mL of the concentrated HNO_3. Then, add enough distilled water to equal 1 L. The resulting solution would be equal to 1 M HNO_3.

If you needed a 3 M solution, multiply the 63 mL by 3 to get 189 mL to add to water to bring it to 1 L. The resulting solution would be a 3 M solution.

To make diluted solutions, chemists use volumetric flasks or beakers for accurate measuring. With practice, making dilutions from concentrated solutions off the shelf will be a snap.

Some people compare chemistry with cooking in the kitchen. Sample preparation, dilution, and mixing are all part of producing a culinary masterpiece. In other words, masterpiece or a mess depends on how you follow the directions.

QUIZ

1. A solution is a homogeneous mixture of a solute in a
   A. colloid.
   B. inert gas.
   C. solvent.
   D. solid.

2. The ability of a compound to dissolve into another is called
   A. vaporization.
   B. condensation.
   C. tensile strength.
   D. solubility.

3. A mole in chemistry
   A. lives underground.
   B. equals the same amount of molecules as the atoms in exactly 12 grams of carbon-12.
   C. seldom contains 10^3 atoms in a sample.
   D. is equal to nearly the number of atoms in exactly 20 grams of carbon-12

4. Molarity
   A. cannot be calculated.
   B. equals the number of moles of solute (n) per volume in liters (V) solution.
   C. never equals concentration.
   D. has the opposite function of dilution.

5. The scattering of light by colloidal-sized particles is called
   A. Avogadro's number.
   B. the Tyndall effect.
   C. a surfactant.
   D. the Taylor effect.

6. Most environmental releases are measured in
   A. parts per hundred.
   B. angstroms per meter.
   C. parts per million.
   D. grams per liter.

7. To dilute a concentrated solution, which of the following are most important?
   A. Molarity and volume
   B. Density and porosity
   C. Volume and acidity
   D. Molarity and area

8. **The sum of the atomic weights of the atoms in a molecule of a substance is known as**

   A. molecular weight.
   B. Avogadro's number.
   C. polar mass.
   D. inverse atomic mass.

9. **When two or more liquids form a solution, they are called**

   A. disparity.
   B. immiscible.
   C. nonpolar.
   D. miscible.

10. **The solubility of a solute is written as**

    A. grams of solvent per 1 g of water at 100°C.
    B. angstroms per kelvin.
    C. grams per liter of mercury.
    D. the ratio of grams of solute per 100 grams of water at a set temperature.

Orbitals?

electron *orbital* is the path it travels around a nucleus like planets around the n. This path determines an electron's role in an atom's properties, especially hemical reactivity. Electrons can be pulled away from atoms, given away by atoms, and even shared among atoms in chemical bonds. In fact, the word orbital was first used by Robert Mulliken, while trying to understand how atoms form bonds. How electrons are sprinkled around the nucleus is the *electron configuration*.

tron Energy Levels

An electron's orbit is created by its energy. Each electron is labeled with a number, letter, and superscript number.

The first label is the *principal quantum number*. It indicates the energy level for the electron called a *shell*. For the first energy level, $n = 1$; second energy level $n = 2$; third energy level $n = 3$; and so on.

The letter following (e.g., 2s) shows the kind of orbit in the shell. The letter s, p, d, or f points out the electron's subshell. Orbitals of the s-type are always singular, p-types form orbital sets of three, d-type orbitals come in sets of five, and f-type orbitals are written in sets of seven. Finally, the superscript shows the number of electrons in a subshell. Figure 7-1 shows the electrons in these

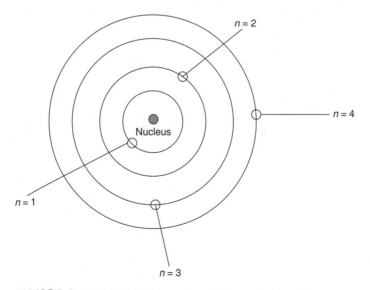

FIGURE 7-1 • Orbital subshell energy levels allow a chemist to figure out bonding of elements.

chapter **7**

Orbitals

Electrons play a big part in the way an atom gets along with others. They d
just bond by accident. Usually, atoms will only combine with their own ato.
or atoms of other elements in certain ways. Electrons are found in atomic an
molecular orbitals depending on whether the electron stays with an atom's
nucleus or is shared between two atoms. In this chapter, you will discover
how valence and electron configuration impacts bonding and molecular
configuration.

CHAPTER OBJECTIVES

In this chapter, you will

- Figure out how an atom's electron configuration relates to bonding
- Learn how orbital filling works
- Understand the Aufbau principle
- Discover what valence shell electron-pair repulsion means in molecular geometry

TABLE 7-1 Electron energy levels have specific capacities	
Quantum number designation (n)	**Shell capacity ($2n^2$)**
1	$2 \times 1^2 = 2$
2	$2 \times 2^2 = 8$
3	$2 \times 3^2 = 18$
4	$2 \times 4^2 = 32$

different orbital subshell energy levels. The overall capacity of the energy levels then looks like Table 7-1.

Spin Magnetism

In 1921, German physicists, Otto Stern and Walter Gerlach found that electrons act like tiny magnets. In their experiments, they used silver atoms and a magnet. Figure 7-2 shows how a beam of atoms sent through a magnet, is split into two beams going in different directions. Stern and Gerlach thought electrons must be attracted and repelled by opposite and like charges just like

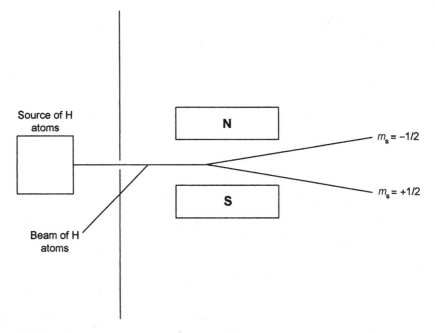

FIGURE 7-2 • Electrons attracted and repelled like magnets.

magnets attract and repel same and like charges. This became known as *spin magnetism* and is written as m_s.

 **EXAMPLE 7-1**

This example shows an orbital diagram of electrons in the s and p orbitals.

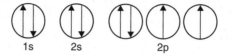

The electrons in orbital diagrams are written as up and down arrows for $\uparrow m_s = +1/2$ and $\downarrow m_s = -1/2$. Think of electrons spinning around orbitals like electrically charged marbles. This electrical charge creates mini magnetic fields within the orbitals.

? Still Struggling

Just as opposite charges attract and like charges repel, electron pairs in orbital subshells are attracted and repelled. If one electron has a value of $m_s = +1/2$, then the other must be $m_s = -1/2$.

Aufbau Principle

The *Aufbau principle* describes electron arrangements of atoms, ions, or molecules. Aufbau is a German word referring to construction, so it's the building of an atom's electrons.

Think of atoms filling subshells like eggs filling a carton. An atom's ground state adds electrons in a specific building order. According to the principle, electrons fill orbitals with the lowest energy *before* filling higher energy orbitals, (e.g., 1s, 2s, etc.).

> The **Aufbau principle** describes how electrons fill the lowest energy orbitals first before filling higher energy orbitals.

The standard building order for an atom's orbital configuration is 1s, 2s, 2p, 3s, 3p, 4s, 3d, 4p, 5s, 4d, 5p, 6s, 4f, 5d, 6p, 7s, and 5f. Figure 7-3 shows the order of an orbital filling sequence. An example of an atom's orbital configuration is illustrated in Fig. 7-4 for the element magnesium.

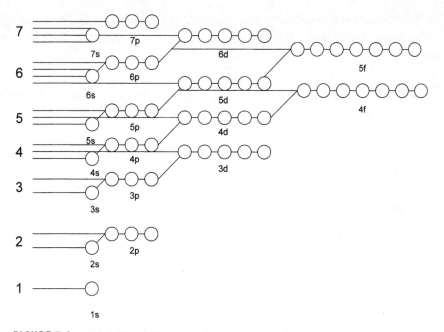

FIGURE 7-3 · Orbital filling follows a regular sequenced order.

Mg

Magnesium

Atomic number – 12

Atomic mass – 24.31

Group – II

Period – 3

Metal

Electrons per orbital layer – 2,8,2

Valence electrons – **$1s^2\ 2s^2p^6\ 3s^2$**

FIGURE 7-4 · The orbital configuration of magnesium shows how it can combine with other elements.

Pauli Exclusion Principle

Before filling those orbitals, however, there are other important rules to know: the Pauli exclusion principle and Hund's rule.

In 1925, Wolfgang Pauli, an Austrian physicist, came up with the *Pauli exclusion principle* based on quantum mechanics. He said no two electrons in an atom could be in the same state or configuration at the same time. This idea was geared toward explaining light emission patterns of atoms. For his discovery, Pauli got the 1945 Nobel Prize in Physics.

> **Pauli exclusion principle** says no two electrons can be in the exact same state or configuration at the same time.

Look at the examples. Are all the orbital diagrams for electrons occupying 1s, 2s, and 3p possible?

EXAMPLE 7-2

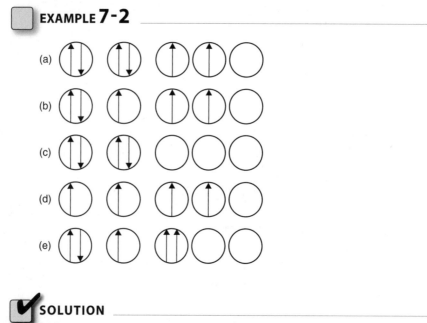

> ### SOLUTION
> Did you get (a) yes, (b) no, (c) yes, (d) no, and (e) no?

Hund's Rule

German physicist Friedrich Hermann Hund helped develop the *molecular orbital theory* while studying the structures of atoms and molecules. During his work, he described characteristics that electrons follow when filling orbitals.

> **Hund's rule** notes that every orbital in a subshell is singly filled with one electron first before any orbital is doubly filled and all electrons in singly-filled orbitals have the same spin.

EXAMPLE 7-3

Carbon has an atomic number of (Z = 6) and orbital configuration of $1s^2 2s^2 2p^2$.

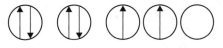

EXAMPLE 7-4

Vanadium, a metal additive to steel, with an atomic number (Z = 23) presents a more complex example. Its orbital configuration is $1s^2\ 2s^2p^6\ 3s^2p^6d^3\ 4s^2$.

? Still Struggling

Hund's rule explains how electrons end up with opposite spins in a shell. No orbital can have two electrons in a subshell until every orbital has at least one electron. The second part of the rule explains that in singly-filled orbitals, all the electrons have the same spin. Otherwise, each electron would end up with opposite spins after pairing.

Subshells and the Periodic Table

Knowing electron configurations of different atoms is helpful in understanding element arrangement and reactivity in the Periodic Table. The groups at the top of the chart, Roman numerals, I–VIII (or alternately 1–18), are used to identify groups and chemical properties as we learned in Chap. 3. These groups show the number of electrons in the outermost orbital of atoms in each vertical column. Groups IA and IIA fill the s subshell. Groups IIIA to VIIIA fill the p subshell. Groups IIIB through VIIIB fill the $(n - 1)$ d subshell. Remember, these outer electrons are known as *valence* electrons.

Just as John Newlands overlapped elements around a cylinder and saw repeating patterns of characteristics, you too can compare element characteristics in the same columns of the Periodic Table. Table 7-2 shows the pattern of elements from the alkali metal family.

TABLE 7-2 There are common properties in the alkali metal family

Element	Melting point (°C)	Boiling point (°C)	Density (g/cm³)
Li	180.5	1347	0.534
Na	97.8	881.4	0.968
K	63.2	765.5	0.856
Rb	39.0	688	1.532
Cs	28.5	705	1.90

Atoms actually contain an infinite number of possible electron configurations. The configuration associated with the lowest energy level is called the *ground state*. When an atom's energy levels go from a lower energy to a higher energy level, the change can be seen as a flash of color or sudden heat.

For example, the electron configuration of calcium ($Z = 20$) in the ground state looks like the following: $1s^2\ 2s^2p^6\ 3s^2p^6\ 4s^2$. It has two electrons in the 1s subshell, eight electrons in the 2sp, eight electrons in the 3sp, and two electrons in the 4s subshell. Another way of writing this is that there are two electrons in the $n = 1$ level, 3 electrons in the $n = 2$ level, eight electrons in the $n = 3$ level, and two electrons in the $n = 4$ level.

EXAMPLE 7-5

Using period and group information from the Periodic Table, what is the configuration of phosphorus, atomic number ($Z = 15$)?

SOLUTION

Start with the first subshell on the Periodic Table as 1s, then in the second period (row) you have 2s. Jumping across in the same row is 2p. In the third period, there is 3s, 3p. In the fourth period there is 4s, 3d, and 4p.

$1s^2$(first period) $2s^2 2p^6$(second period) $3s^2 3p^3$ (third period)

Phosphorus is in the third period so $n = 3$ and group 5A so valence electrons = 5.

(Remember, the valence electron shell arrangement is the same as the outermost placement of electrons.)

EXAMPLE 7-6

Try nickel with atomic number ($Z = 28$).

SOLUTION

First consult the Periodic Table for the period number and group. Then start building the subshells. Did you get the following:

$$1s^2 2s^2 2p^6 3s^2 3p^6 4s^2 3d^8$$

EXAMPLE 7-7

If cesium ($Z = 55$) has a configuration of

$$1s^2 \ 2s^2 p^6 \ 3s^2 p^6 d^{10} \ 4s^2 p^6 d^{10} \ 5s^2 p^6 \ 6s^1$$

What is its period number and group? What is the filling valence subshell?

SOLUTION

Cesium is an alkali metal in sixth period and group 1. The filling orbital is $6s^1$.

EXAMPLE 7-8

Take a look at oxygen ($Z = 8$). The electron configuration of oxygen is

$$1s^2 2s^2 p^4$$

Two electrons in the 1s subshell, two electrons in the 2s, and four electrons in the p subshell give a total of eight.

EXAMPLE 7-9

The electron configuration of boron ($Z = 5$) contains two electrons in the 1s subshell, two electrons in the 2s subshell, and one electron in the 2p subshell. It is written as follows:

$$1s^2 2s^2 2p^1$$

Elements of the same group have identical valence electron structures, when considering numbers of paired and unpaired valence electrons. These similar valence structures cause family group members to react similarly.

Additionally, elements with a lot of orbitals far from the nucleus are increasingly more reactive as the electrons zip through a larger area. With more "party room" to come in contact with other atoms, they take advantage of it!

Ionization Energy

Valence electrons are far from the nucleus and more likely to be pulled away or shared between elements. The only exceptions are the ideal gases, because their outer orbitals are full of electrons already.

When an electron is lost, the balance between electrons and protons is also lost. Without an electron's negative charge, an atom becomes ionized. With more protons than electrons, an ion carries a positive charge. The amount of energy needed to pull away an electron is called the *ionization energy*. The more electrons an element has, the harder it is to pull off an electron.

In fact, ionization energy increases from left to right within the same period and if you look at elements from left to right along any row of the Periodic Table, you'll see regular patterns.

> **Ionization energy** of an element is the energy needed to grab an electron from an element.

Differences in an element's properties are related to changes in electron arrangement. The higher the number of electrons in an atom's outermost shell, the higher an atom's ionization energy.

Elements have been placed in rows within the Periodic Table to make it easier to use. For example, if you are interested in a specific element, check out its place in the Periodic Table. Who are its neighbors? Which group is it in? Which period? How many electrons are in its outmost orbit (i.e., outermost shell)? Is it reactive or not? Is it a metal or nonmetal?

Figure 7-5 illustrates a beryllium atom with its energy levels. A beryllium atom is made up of four protons, five neutrons in the nucleus, and four electrons in two orbital shells outside the nucleus. The first and second shells each have two electrons.

Check out Fig. 7-6. All the group and period information you'll need on the elements can be found on a Periodic Table.

Knowing the reactivity of an element is important. If you are studying potassium and put it into water, you would have a wild reaction because alkali

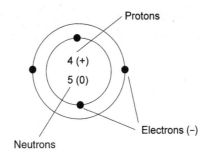

Beryllium (Be)

FIGURE 7-5 • Beryllium atom with its energy levels.

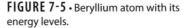

FIGURE 7-6 • Periodic Tables contain group and period information.

metals get really violent in water. They give off hydrogen gas that ignites with the heat of the reaction and gives off a violet flame, like mini-fireworks. From potassium's soft solid state and its low boiling point, you might think it is a mild-mannered element. However, the heat caused by its interaction with water changes a solid chunk of potassium into a liquid by melting it. Can't judge an element by its appearance!

Valence Bond Theory

In 1916, G. N. Lewis, an American physical chemist, proposed that two unshared electrons between atoms could interact to form a chemical bond. These bonding electrons came from unpaired electrons filling each atom's outermost orbital. The orbitals overlapped to form a bond with the electrons between the atoms.

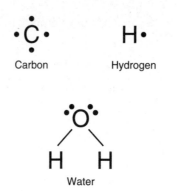

FIGURE 7-7 • Lewis electron-dot structures show the location of the unpaired electrons.

Then in 1930, American chemist, Linus Pauling expanded Lewis's idea by describing the *hydridization* (or mixture) of orbital and resonance theory. The work of Lewis, Pauling, and others created what was eventually called the *valence bond theory*.

Lewis Electron-Dot Structures

Lewis came up with an easy way to draw bonded atoms and explain his theory. He found that if valence electrons are written as dots around a central atom, unpaired electrons are easy to spot. Electron pairing between two atoms forms a bond containing both electrons. Lewis dot structures are a simple way to show bonds and the formation of ions. Figure 7-7 shows examples of Lewis electron-dot structures.

Molecular Orbital Theory

Along with valence bond theory, Friedrich Hund and several others came up with another explanation for atomic bonding called the *molecular orbital theory*. It explains chemical bonding in a way that borrows a lot from the mathematics of wave functions used in quantum chemistry.

Molecular orbital theory uses a row of combined atomic orbitals to form bonds between atoms. The bond is shown with a molecular orbital covering the entire molecule. The electron is more often located between the atoms instead of elsewhere. When it is mostly between the atoms, a bond forms.

This idea is different from valence bond theory, which limits electron placement to between atoms. Although first thought to be at odds, valence bond theory and molecular bond theory are complementary bonding ideas with different strengths and weaknesses.

In molecular orbital theory, orbital volume surrounds the entire molecule. Each arrow ($\uparrow\downarrow$) in a molecular orbital diagram represents a molecular orbital. This is the area where a higher percentage of negative charge, created by the electron, is found. It is thought that electrons fill molecular orbitals of molecules like electrons fill atomic orbitals in atoms. Specifically:

1. Molecular orbitals are filled yielding the molecule's lowest potential energy.
2. Only two electrons are allowed per molecular orbital (i.e., Pauli's exclusion principle).
3. Equal energy orbitals fill with parallel spin before pairing up (i.e., Hund's rule).

Linear Combination of Atomic Orbitals

In the field of quantum mechanics, electron configurations of atoms are calculated as wave functions. In chemical reactions, orbital wave functions are affected (i.e., electron cloud shape changes) according to the type of atoms making up the chemical bond. *Linear combination of atomic orbitals* (LCAO) is a quantum chemistry method for calculating molecular orbitals based on their phase relation.

Molecular orbital diagrams describe the bonding of overlapping electrons. These commonly take place when:

- Molecular orbitals are formed at the overlap of atomic orbitals
- Atomic orbitals of nearly the same energy combine
- Two atomic orbitals overlap and interact to form two molecular orbitals (i.e., in-phase bonding and out-of-phase bonding)

Quantum chemistry is beyond the scope of an introductory chemistry book, but it is an interesting discipline to explore during future studies.

Resonance Theory

When structures can't be easily represented by a single electron-dot structure, but as intermediates between two or more structures, it is known as *resonance*. In valence bond theory, bonds between two atoms may alternate between single and double bonds to result in a combination of both. In molecular orbital

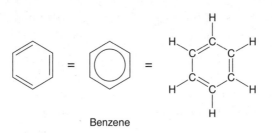

Benzene

FIGURE 7-8 · Benzene forms bonds at several locations.

theory, the bond takes the form of an orbital encompassing the whole molecule and predicts single and double bonds possibilities.

Benzene (C_6H_6), a colorless, flammable liquid with a sweet odor, is a volatile chemical that evaporates quickly. Benzene exhibits resonance theory since bonding is going on constantly between the double-bonded carbons. Figure 7-8 shows benzene with bonds in various locations and with shared bonds (i.e., drawn as a circle within the hexagonal structure). Resonance molecules are often mixtures of different forms in stable configurations.

A molecule's written configuration gives only a percentage of possible configurations, since there is usually a mingling of arrangements. In fact, at the atomic level molecular shapes change so fast that specific locations and momentum of individual electrons aren't exactly known. This condition is a description of *Heisenberg's uncertainty principle.*

In 1927, Werner Heisenberg, a German physicist, suggested it was impossible to know both the location *and* speed of anything in a specific point of time with certainty. Specific attention is on one or the other. In this case, his principle is commonly used to describe electrons. Heisenberg won the 1932 Nobel Prize in Physics for this uncertainty principle.

> **Heisenberg's uncertainty principle** notes that it is impossible to identify a particle's exact momentum and position at the same moment in time.

Sigma and Pi Bonds

Regardless of bonding theory, overlapping orbitals form two common types of bonds: sigma and pi bonds. *Sigma bonds* involve a head-on-head overlap between orbitals. An s orbital is a sphere, so that two s orbitals overlap to form a sigma bond. Single bonds are always sigma bonds.

Pi bonds happen when two orbitals are in parallel. Two bonds between two atoms, or a double bond, consist of one sigma and one pi bond each. Three

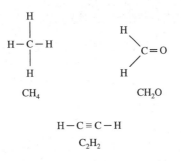

FIGURE 7-9 · Stable, carbon-based molecules can have single, double, and triple bonds.

bonds between two atoms, or a triple bond, consist of one sigma and two pi bonds. Carbon likes to form all these bonds. When it does, they are known as organic compounds and are described in Chap. 12. Figure 7-9 shows examples of carbon-based molecules with single, double, and triple bonds. Energetically, sigma bonds are much stronger than pi bonds.

Molecular Geometry

Structure always determines function. A molecule's shape or geometry is important for discovering how it interacts with itself and other molecules. When atoms combine into a simple molecule, they take positions with their electron pairs as far apart as possible.

Valence Shell Electron-Pair Repulsion (VSEPR)

A molecule's shape is affected by the location of the electron pairs repelling one another. It is a bit like a middle school dance, where boys and girls tend to stay far apart. They may briefly move closer, but spin away again. VSEPR is the same idea. Molecular geometries in space are affected by the nearness of electron pairs to each other. Bond angles reflect this positioning.

Two Electron Pairs (Linear)

An atom with two electron pairs such as calcium chloride $(CaCl_2)$ is linear $(180°$ bond angle). Carbon dioxide (CO_2) is also linear, but its Lewis structure contains two double bonds. Commonly multiple bonds should be treated as a group of electron pairs in molecular geometry.

Three Electron Pairs (Trigonal Planar)

The geometry of a molecule with a central atom containing three pairs of electrons is *trigonal planar*. BF_3 has this configuration. Then, if a bonding pair with a lone pair is replaced (i.e., SO_2), the geometry becomes bent or angular.

Water Geometry

Water (H_2O) molecules have a bent molecular geometry shaped like a boomerang with oxygen at the bend. Oxygen has six valence electrons and needs two more electrons from two hydrogen atoms for a valence shell octet. With two bonds to different hydrogen atoms, there are two lone unbonded electron pairs. The two hydrogen atoms and the two lone electron pairs are positioned far apart at nearly a 109° bond angle or *tetrahedral electron pair* geometry. Then, the two lone electron pairs push the shape even further apart from the two bonding hydrogen atoms to form a slight bend and a 104° bond angle.

Four Electron Pairs (Tetrahedral)

The geometry of a molecule with four electron pairs, like carbon tetrachloride (CCl_4), is tetrahedral (109.5° bond angle). For example, when bonding pairs are replaced with nonbonding pairs, the molecular geometry can become *trigonal pyramidal* (three bonding and one nonbonding), bent or angular (two bonding and two nonbonding), and linear (one bonding and three nonbonding).

H_2O and SO_2 have similar bent geometry, but their bond angles are different. For H_2O the H-O-H angle is around 104°, while in SO_2 the O-S-O angle is about 118°.

Five Electron Pairs (Trigonal Bipyramidal)

A molecule with a central atom containing five pairs of electrons is *trigonal bipyramidal* (e.g., PCl_5). When bonding pairs are replaced with nonbonding pairs, the molecular geometry adjusts to seesaw (four bonding and one nonbonding), T-shaped (three bonding and two nonbonding), and linear (two bonding and three nonbonding). This moves the end atoms into axial and equatorial locations.

Six Electron Pairs (Octahedral)

Geometry for a molecule with a central atom containing six electron pairs is octahedral, such as SF_6. When bonding pairs are replaced with nonbonding pairs, molecular geometry changes from square pyramidal (five bonding and one nonbonding) to square planar (four bonding and two nonbonding). No other bonding combinations work in octahedral electron-pair geometry.

QUIZ

1. **In 1921, Stern and Gerlach discovered spin magnetism where**
   A. elements always change from solid to liquid within a few degrees.
   B. electrons are only attracted by like charges.
   C. electrons are attracted and repelled by opposite and like charges, respectively.
   D. freezing points are proportionally decreased.

2. **Element groups**
   A. give the number of protons in the single outermost orbits.
   B. give the number of electrons in the outermost orbitals of atoms in each column.
   C. are rarely used in figuring out reactivity.
   D. are arranged by color.

3. **An element's ionization energy**
   A. cannot be calculated.
   B. creates mini magnetic fields.
   C. is the inverse of inertial energy.
   D. is the energy needed to detach an electron from an elemental atom.

4. **Ground state is**
   A. the electron configuration associated with an element's lowest energy.
   B. only applied to heavy metals.
   C. the energy of an element's outermost orbital levels.
   D. not related to nobility.

5. **What describes how all orbitals of a given sublevel must be occupied by a single electron before pairing begins?**
   A. Hunt's rule
   B. Pauli's rule
   C. Hund's rule
   D. Palmer's rule

6. **Elements in column II of the Periodic Table have**
   A. two free electrons in the outermost orbit around the nucleus.
   B. three free electrons in the outermost orbit around the nucleus.
   C. four free electrons in the outermost orbit around the nucleus.
   D. five free electrons in the outermost orbit around the nucleus.

7. **An atom's electron configuration is**
   A. found only when performing mass balance equations.
   B. determined by the kinetic energy present.
   C. found by calculating melting point.
   D. written as s, p, d, and f subshells.

8. The Pauli exclusion principle states that
   A. any elements with a free d orbital can bond.
   B. no two electrons in the same atom can be in the same configuration at the same time.
   C. two atoms sharing an orbital are matched exactly.
   D. electron attraction in a magnetic field is a result of paired electrons.

9. How many electrons may occupy an orbital at the same time?
   A. 1
   B. 2
   C. 3
   D. 4

10. If the number of electrons an energy level can hold is $2n^2$, n is
    A. the number of neutrons found at ground state.
    B. dependent on the atom's specific gravity.
    C. the principal quantum number and an electron's energy level.
    D. the normalized temperature.

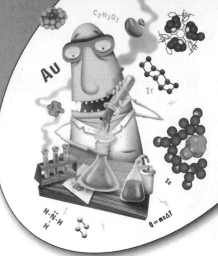

chapter **8**

Chemical Bonds

In this chapter we will learn the ins and outs of chemical bonding. We will also learn how some bonds are formed when atoms or molecules share electrons, while others are formed between charged atoms or molecules after gaining or losing electrons.

CHAPTER OBJECTIVES

In this chapter, you will

- Familiarize yourself with electronegativity
- Discover the different kinds of covalent bonds
- Understand dipole moment
- Learn all about ions and ionic bonds
- Find out about two different compound naming systems

What Are Covalent Bonds?

All elements are neutral because of an equal number of electrons and protons, but the elements are not usually stable. As discussed in Chap. 7, some electron arrangements take place more often than others. In the simplest case, atoms obey the octet rule. For larger elements, different electron arrangements are more common due to the different electron orbital types possible.

To pull off the best electron arrangements, atoms have two choices. They can share their electrons with other atoms to form *covalent bonds* or give up and collect electrons from other atoms to make ions. Even though electrons are moved around, ions need partners to make compounds. Those interactions involve *ionic bonds*. Ideal gases are the only exception to these types of interactions, since they already have the perfect electron arrangement.

> **Covalent bonds** form when atoms or molecules share electrons, whereas **ionic bonds** form between two charged atoms or molecules created after gaining or losing electrons.

Electronegativity

Whether electrons share electrons or not depends on an atom's *electronegativity*. Electronegativity is a chemical property which describes an atom or group's ability to attract electrons. Conversely, *electropositivity* is the tendency to give up electrons to other atoms or molecules.

Linus Pauling discovered electronegativity while studying the electrical charge of molecules. He noticed atoms and molecules have different abilities to give up or accept electrons.

> **Electronegativity** is the ability of an atom or group to attract electrons, while **electropositivity** is the ability of an atom or group to give up electrons to other atoms or groups.

To explain his observations, Pauling developed a scale for electronegativity. He found that fluorine was the most electronegative atom and could easily pick up an extra electron during a reaction. He gave it an electronegativity number

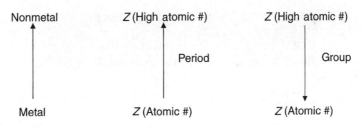

FIGURE 8-1 · Electronegativity values can be compared with atomic number.

of 4.0. Weakly electronegative atoms (e.g., alkali metals) have values around 0.7 on the Pauling electronegativity scale.

Still Struggling

Atoms or molecules at the extremes of the Pauling scale are most likely to gain or lose electrons. High electronegativity favors electron gain, while low electronegativity favors electron loss. Middle-the-scale atoms or molecules are more likely to share electrons via covalent bonds. Figure 8-1 shows the way electronegativity values increase compared to atomic number.

EXAMPLE 8-1

Look at the following example. Using the values in Table 8-1, can you find the electronegativity values of tellurium (Te) and calcium (Ca)?

$$Te = atomic\ number\ (Z) = 52$$
$$Ca = atomic\ number\ (Z) = 20$$

Which element has the higher electronegative value? Did you get tellurium? How about the electronegativity order of mercury, oxygen, and beryllium from the highest value to the lowest?

SOLUTION

$$Br = atomic\ number\ (Z) = 35,\ O = atomic\ number\ (Z) = 8,$$
$$Hg = atomic\ number\ (Z) = 80$$

Did you get oxygen, bromine, and mercury?

TABLE 8-1	The Pauling scale of electronegativity helps to find an element's bonding potential	
Element	Z	Electronegativity value
Fr	87	0.7
Ca	20	1.0
Hf	72	1.3
Be	4	1.5
Co	27	1.8
Ge	32	1.8
Hg	80	1.9
Re	75	1.9
Sb	51	1.9
Te	52	2.1
At	85	2.2
Br	35	2.8
N	7	3.0
O	8	3.5

Covalent Compounds

Covalent bonds take place through electron sharing between two atoms. The strength of these covalent bonds depends on the electronegativity between the bonded atoms and the bonding angle.

There are many types of covalent bonds. Covalent bonds may form between nonmetals and metals. However, the most common type of covalent bond is formed between nonmetals. *Covalent compounds* are generally soft solids with low melting points. In fact, many are liquids or gases at room temperature. When only two elements are bonded, the compound is called a *binary covalent compound*.

Think of two college friends who want to move a heavy television weighing 200 lb. into their dormitory suite. Mark can lift 75 lb. easily. His friend, Mike, works out at the gym every day, is a collegiate discus thrower, and can lift 125 lb. Neither student can lift and move the television by themselves, but together they can lift 200 lb. When combining their abilities, they are able to move the television into their dorm suite. If either drops a side, it will be a big problem!

The way different compounds approach bonding is like the roommate example. By bonding with like atoms, compounds get the best electron arrangement. If either compound breaks the bond, the goal (e.g., product formation) isn't possible.

EXAMPLE 8-2

The covalent bond between hydrogen is simpler than ionic bonding

$$H + H \rightarrow H : H \text{ or } H_2$$

The shared electron pair in a molecule is known as a *covalent bond* or a *covalent pair*. It is a jointly useful connection.

Polarity

Covalent bonds are characterized by electron sharing, but the sharing doesn't have to be equal. When atoms share electrons equally, the covalent bond is said to be *nonpolar*. When electrons are not shared equally between atoms, the covalent bond is *polar*. Electronegativity between bonded atoms makes a covalent bond polar or nonpolar.

> A covalent bond is **polar** when the electrons are not shared equally between the atoms and **nonpolar** when electrons are equally shared.

Nonpolar Covalent Bonds

Nonpolar bonds form between atoms of same or similar electronegativities. These bonds form between like atoms, as well as different ones. Diatomic compounds, such as H_2 and I_2, are good examples of nonpolar bonded molecules. Hydrogen and iodine have very different electronegativities but when they bind to the same kind of atom, they share electrons equally. They are like identical twins.

Let's take a look at bonds between different atoms. Carbon and hydrogen, with similar electronegativities, bond to other carbons as well as hydrogen. While carbon and hydrogen can only form a single bond, carbon can form single, double, or even triple bonds as long as the total bond count is four. (Look back to Fig. 7-9 for examples.)

The study of carbon compounds is so important that an entire branch of chemistry (i.e., organic chemistry) focuses on it. In Chap. 12 you will learn much more about organic chemistry.

Whether a bond forms between the same or different atoms, however, the resulting nonpolar bonds are the strongest covalent bonds, because the electrons are shared fairly equally between the atoms.

Polar Covalent Bonds

Unlike nonpolar bonds, polar covalent bonds form between atoms with different electronegativities. Unequal electron sharing is why the bonds are considered polar. This means electrons spend more time at one atom than the other. Because electrons are negatively charged, the atom where they spend the most time has a partial negative charge. The other bonded atom gets the shared electrons less and so it will be partially positive. This unequal electron sharing makes polar bonds weaker than nonpolar bonds. In fact, it impacts important properties of polar compounds such as the ability to form and break hydrogen bonds. More on this when we study acids and bases in Chap. 10.

Dipole Moment

For nonpolar bonds, there is no charge separation (i.e., sharing is equal) and no dipole moment exists. But for polar bonds, charge separation causes polarity. In fact, the dipole strength affects molecular properties including chemical reactivity and solubility.

> A **dipole moment** is the measurement of charge separation between each part of a molecule.

Dipole arrangements in a molecule also impact molecular behavior. When they are pointing in opposite directions, dipoles cancel each other out. This arrangement, found in molecular geometries such as linear, equilateral triangle, or tetrahedron, was described in Chap. 7.

A good example of dipole effects can be seen when comparing carbon tetrachloride (CCl_4) and chloroform ($CHCl_3$). The bond formed between carbon and chlorine is particularly polar based on very different electronegativities. For carbon tetrachloride, all four bonds point in opposite directions, so the dipoles cancel out. In other words, carbon tetrachloride is *nonpolar*, even though it is made up of polar bonds.

This is not the case for chloroform. The presence of the bond between carbon and hydrogen means there is a dipole imbalance, so chloroform is polar.

In fact, calculating dipole moments in advanced chemical equations, is an important tool for chemists to use in deciphering element interactions. The dipole moment is commonly measured in *debye* units (D). There is also an SI unit for the dipole moment, the coulomb-meters (C·m). The relationship between these units is as follows:

$$1\ D = 3.34 \times 10^{30}\ C \cdot m$$

Water (H_2O) is a very polar molecule. An extremely electronegative oxygen atom pulls electrons from hydrogen in the covalent bond between the two atoms. This separation of charge creates a dipole moment equal to 1.94 D or 6.48×10^{30} C·m.

Naming Covalent Compounds

When naming a covalent compound, the kind and number of atoms in the compound are used. To name a binary covalent compound, the IUPAC (*International Union of Pure and Applied Chemistry*) method orders the nonmetals in a certain way. The ordering of the nonmetals is as follows:

B > Si > C > P > N > H > S > I > Br > Cl > F > O > functional group

The root of the first nonmetal is the same as the element. For the second nonmetal, the end of the word is changed to "*ide*" in most cases. Lastly, a Greek prefix is included to indicate the number of atoms as shown in Table 8-2.

For example, BF_3 contains one boron and three fluorine atoms and is written with boron *before* the fluorine atoms. Based on the IUPAC rules, (BF_3) is named boron trifluoride. In time, naming compounds will become second nature.

EXAMPLE 8-3

Name the following compounds: CO, CO_2, N_2O_3, CCl_4, SF_6.

SOLUTION

Did you get carbon monoxide, carbon dioxide, dinitrogen trioxide, carbon tetrachloride, and sulfur hexafluoride?

TABLE 8-2	Greek prefixes and counting atoms
Number	Greek prefix
1	Mono
2	Di
3	Tri
4	Tetra
5	Penta
6	Hexa
7	Hepta
8	Octa
9	Nona
10	Deca
12	Dodeca

What Are Ions?

If the sharing of electrons is not possible, an atom or molecule will gain or lose electrons and no longer remain neutral. During this process, called *ionization*, charged *ions* are formed.

Usually, metals lose electrons and nonmetals gain electrons.

> Metals ⬇ electrons and form cations
>
> Nonmetals ⬆ electrons and form anions

The charge results from a change in the amount of negatively charged electrons and positively charged protons. If an atom or molecule gains an electron, a negatively charged *anion* is formed. The extra electron makes anions larger than the original atom.

If an atom or molecule loses an electron, a *cation* is formed. The loss of an electron makes the cation smaller than the original atom. Like covalent bonds, the driving force behind electron activity within atoms is getting the most stable arrangement of electrons. Smaller atoms try to fulfill the octet rule, while larger atoms are a bit tricky. Larger atoms do not always obey the octet rule, since there are a lot more orbitals to fill.

Additionally, there are also preferred cation and anions forms. This is common for transition metals like chromium, iron, tin, and copper. In the case of these molecules, one atom wants to gain or lose electrons, but the resulting change affects the entire molecule.

Types of Ions

Monatomic ions consist of one or more atoms of the same element. The "mon" part of the name indicates there is only one type of atom present. Monatomic ions can be either anions or cations. Examples of anions include chloride (Cl^-), sulfide (S^{2-}), and nitride (N^{3-}), while sodium (Na^+), magnesium (Mg^{2+}), and barium (Ba^{2+}) are common cations.

> **Monatomic ions** are charged species that consist of one or more atoms of the same element.

Some compounds can exist in nature at different ionic states, such as copper (Cu^+, Cu^{2+}), iron (Fe^{2+}, Fe^{3+}), and gold (Au^+, Au^{3+}).

The name "poly" indicates more than one atom type is present. The element name is taken from the innermost or core atom. For example, the ammonium ion (NH_4^+) is a common polyatomic ion.

A majority of *polyatomic ions* involve elements bonded with oxygen. Since oxygen is critical to biological systems, polyatomic ions are key components. The most important polyatomic cation is the hydronium ion (H_3O^+), which will be discussed in the next section. Table 8-3 gives a few monatomic and polyatomic cations (+) and their charges.

> **Polyatomic ions are** charged species that contain two or more different elemental atoms.

Most of the common inorganic (no carbon) polyatomic ions have negative charges, such as arsenite (AsO_3^{3-}), a water toxin, and phosphate (PO_4^{3-}), an important biological compound.

Organic or carbon-based compounds can also exist as anions. Oxalate ion ($C_2O_4^{2-}$) is an organic anion found in clover and coffee, which hampers iron

TABLE 8-3 Monatomic and polyatomic cations are important to keep track of in reactions

+1	+2	+3 and +4
Hydrogen ion H^+	Barium ion Ba^{2+}	Aluminum ion Al^{3+}
Lithium ion Li^+	Beryllium ion Be^{2+}	Gold(III) Au^{3+}(auric)
Sodium ion Na^+	Calcium ion Ca^{2+}	Cobalt(III) Co^{3+} (cobaltic)
Potassium ion K^+	Cobalt(II) Co^{2+} (cobaltous)	Chromium(III) Cr^{3+} (chromic)
Rubidium ion Rb^+	Chromium(II) Cr^{2+} (chromous)	Iron(III) Fe^{3+}(ferric)
Cesium ion Cs^+	Copper(II) Cu^{2+} (cupric)	Lead(IV) Pb^{4+}(plumbic)
Silver ion Ag^+	Iron(II) Fe^{2+}(ferrous)	Tin(IV) Sn^{4+}(stannic)
Copper(I) Cu^+ (cuprous)	Mercury(II) Hg^{2+} (mercuric)	
Gold(I) Au^+ (aurous)	Mercury(I) Hg_2^{2+} (mercurous)	
	Magnesium ion Mg^{2+}	
	Nickel ion Ni^{2+}	
	Radium ion Ra^{2+}	
	Tin(II) Sn^{2+}(stannous)	
	Strontium ion Sr^{2+}	
	Zinc ion Zn^{2+}	

absorption in the intestine. Table 8-4 provides a few commonly used monatomic and polyatomic anions (−).

Ions and Hydrogen

Ions are also formed through covalent bonding with hydrogen. Hydrogen bonds can be very weak, so they are frequently broken and reformed in many different ways. This process often leads to changes in a molecule's charge. While the topic is mentioned here, Chap. 10 on acids and bases describes this important chemical process in greater detail.

When hydrogen loses an electron, only a positively charged proton is left. Although a proton is a cation, protons never exist by themselves in solution.

TABLE 8-4 Monatomic and polyatomic anions are important to keep track of in reactions		
−1	**−2**	**−3 and −4**
Acetate $C_2H_3O_2^-$	Carbonate CO_3^{2-}	Arsenide As^{3-}
Acetate CH_3COO^-	Chromate CrO_4^{2-}	Borate BO^{3-}
Bromate BrO_3^-	Dichromate $Cr_2O_7^{2-}$	Carbide C^{4-}
Bromide Br^-	Hydrogen phosphate HPO_4^{2-}	Nitride N^{3-}
Chlorate ClO_3^-	Oxalate $C_2O_4^{2-}$	Phosphate PO_4^{3-}
Chloride Cl^-	Oxide O^{2-}	Phosphide P^{3-}
Cyanide CN^-	Selenide Se^{2-}	Phosphite PO_3^{2-}
Hydride H^-	Sulfate SO_4^{2-}	
Hydrogen carbonate HCO_3^-	Sulfide S^{2-}	
Hydrogen sulfate HSO_4^-	Telluride Te^{2-}	
Hydroxide OH^-	Thiosulfate $S_2O_3^{2-}$	
Iodide I^-		
Nitrate NO_3^-		
Perchlorate ClO_4^-		
Permanganate MnO_4^-		
Thiocyanate SCN^-		

They always have to partner up with another atom to get a pair of electrons. Hydrogen must fill its lonely empty orbital. In water, there's a lot of H_2O everywhere. H_2O is a good partner for a proton. Its oxygen atom can share a pair of electrons with a proton to complete its lower orbital (see Chap. 7). Because a proton has a positive charge, the resulting molecule, hydronium ion H_3O^+, gains a positive charge. Other atoms and parts of molecules can partner up with protons to create ions as well [e.g., ammonia (NH_3) to form an ammonium ion (NH_4^+)].

In the same way, a proton can be stripped from a neutral molecule to create an anion. The most common example is water (H_2O) losing a proton (H^+) to make hydroxide ion (OH^-). This time, the proton loss leaves an extra electron with oxygen so it keeps a pair of electrons. The resulting hydroxide ion has one more electron than total protons, and a full negative charge. Oxygen doesn't mind having an extra electron because of its electronegativity. Other neutral molecules gain a negative charge by losing a proton, such as carbonic acid (H_2CO_3), which forms bicarbonate (HCO_3^-) in baking soda (sodium bicarbonate).

Ionic Bonds

The old saying, "opposites attract" applies to ions! Anions and cations interact to form strong *ionic bonds*. Ionic compounds are hard crystalline solids with high melting points due to their strong ionic bonds. The ions' opposite charges attract one another through *electrostatic interactions*.

Cations and anions mix by creating and breaking ionic bonds. So, one way to mix ions is by making a solution. Although ionic compounds form strong bonds, water has the ability to surround ions and separate them. This process, known as *solvation* makes ions into electrolytes, which conduct electricity (see Chap. 9).

Balancing Ionic Compounds

Ionic bonds form between charged species so that the compound's overall charge is neutral. When both ions have a charge of 1, like sodium (Na^+) and chloride (Cl^-), the compound adds them together to make NaCl.

If charges are different, the smallest number of each are combined to make the final compound. Magnesium (Mg^{2+}) and chloride (Cl^-) make magnesium chloride $(MgCl_2)$, so that the two $(-)$ chlorides cancel the two $(+)$ charges on the magnesium ion.

Ammonium carbonate $[(NH_4)_2CO_3]$ is a similar example. In this case, a parentheses is used to show that two ions of ammonium are needed to balance one carbonate ion.

One strategy is to find the *least common multiplier* (LCM) when balancing the parts of a molecule. The LCM is the smallest positive number that is a multiple of two numbers. In the previous example, the numbers on the ions for magnesium chloride and ammonium carbonate were 1 and 2, so that the LCM is 2. But what if the numbers on the ions are 2 and 3?

Gadolinium oxide is used to make a contrast agent for *magnetic resonance imaging* (MRIs). It is made from Ga^{3+} and O^{2-} ions. What is the LCM for gadolinium oxide? Can you figure out the number of ions in the compound? Neither number can be divided by the other without getting a remainder. The LCM (6) is found when the numbers are multiplied by one another.

For balancing the charges on a molecule, there need to be two gadolinium ions with a charge of 3, and there need to be three oxide ions with a charge of 2. In both cases, multiplying the charges by the number of ions gets the same LCM of 6. The final formula for gadolinium oxide is then Ga_2O_3. With practice, balancing ions will soon become fairly straightforward.

Naming Ionic Compounds

To name an ionic compound, name the ions within the compound. The naming order is cation first and anion second. In most instances, a metal cation is named first, followed by a nonmetal anion.

When a cation comes from an element, the name doesn't change, like sodium (Na^+), magnesium (Mg^{2+}), and calcium (Ca^{2+}). If an anion comes from an element, the root of the name is the element with "*ide*" added at the end, such as hydr*ide* (H^-), chlor*ide* (Cl^-), and nitr*ide* (N_3^-). When combined, the named compounds become sodium hydr*ide* (NaH), magnesium chlor*ide* ($MgCl_2$), and calcium nitr*ide* [$Ca(N_3)_2$].

In some cases, this strategy doesn't work. Some elements form more than one ion, so two naming methods are used for those compounds: the Common Naming System and the newer Stock System.

The Common Naming System

Common names don't often follow the rules, but they can when naming ionic compounds. Ionic compounds use a naming prefix for the element and a suffix to describe the specific ion present. This compound naming method is used for cations and anions.

When metal cations exist in two states, "*ous*" and "*ic*" are added as suffixes, respectively. For example, if an experiment calls for cupr*ous* sulfate, it is talking about the copper ion (Cu^+) and not (Cu^{2+}), which is called a cupr*ic* ion. Similarly, iron exists as ferr*ous* (Fe^{2+}) and ferr*ic* (Fe^{3+}) ions. In these two examples, the suffixes do not indicate the same charge on the ion, but only whether it is the lower or higher numbered one. In other words, you still need to know what two forms of the ions are possible.

When nonmetal polyatomic anions exist in multiple states, a series of suffixes are used to describe all possible compounds. When there are only two types, the "*ate*" and "*ite*" ending are used. The polyatomic ion with the most atoms is called "ate," while the least is called "ite."

A sulfur atom attached to four oxygen atoms is a sulf*ate* ion (SO_4^{2-}). A sulfur atom attached to three oxygen atoms is a sulf*ite* ion (SO_3^{2-}). In some cases, there

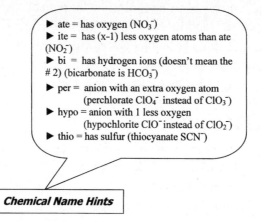

▶ ate = has oxygen (NO_3^-)
▶ ite = has (x-1) less oxygen atoms than ate (NO_2^-)
▶ bi = has hydrogen ions (doesn't mean the # 2) (bicarbonate is HCO_3^-)
▶ per = anion with an extra oxygen atom (perchlorate ClO_4^- instead of ClO_3^-)
▶ hypo = anion with 1 less oxygen (hypochlorite ClO^- instead of ClO_2^-)
▶ thio = has sulfur (thiocyanate SCN^-)

Chemical Name Hints

FIGURE 8-2 • A few clues to chemical naming can make bonding a lot easier to understand.

are four different types of polyatomic anions. These are the guidelines for naming these compounds:

- An ion with four other atoms is named: "per___ate"

- An ion with three other atoms is named: "___ate"

- An ion with two other atoms is named: "___ite"

- An ion with one other atom is named: "hypo___ite"

For polyatomic bromine anions, the names would be *per*brom*ate* (BrO_4^-), brom*ate* (BrO_3^-), brom*ite* (BrO_2^-), and *hypo*brom*ite*, (BrO^-).

Figure 8-2 gives a few clues to help with chemical naming.

As you've learned, compounds' characteristics are affected by their ionic character. It is easier to understand and remember element bonding if you memorize or make flash cards to study these ionic forms and their types.

Stock System

Alfred Stock, a German chemist at Cornell University, wanted to improve the names of ionic compounds to better describe their makeup. In 1919, he published the *Stock System*. This new naming system for compounds shows the ion charge with a Roman numeral. There is no guessing about the charge this way. The common name for $FeCl_2$ is ferrous chloride, because it contains Fe^{2+} and not Fe^{3+}. Based on the Stock System, however, $FeCl_2$ is called iron(II) chloride. Ferric chloride ($FeCl_3$) is iron(III) chloride. The IUPAC naming system uses the Stock system, but both systems are still used.

QUIZ

1. What is the overall charge of a phosphate ion, PO_4^{3-}?

   A. Positive charge
   B. No charge
   C. Has no electromagnetic property
   D. Negative charge

2. When writing formulas for polyatomic ions, you multiply each charge by whatever number works to give the

   A. pH.
   B. electronegativity number.
   C. lowest common multiple.
   D. highest common multiple.

3. What is formed when an electron is stripped from an atom?

   A. Anion
   B. Cation
   C. Noble gas
   D. Heavy metal

4. Which is the name for (BrO_4^-)?

   A. Perbromate
   B. Bromic oxide
   C. Bromate
   D. Hypobromite

5. Covalent bonds share electrons, while ionic bonds

   A. are rarely formed.
   B. are much more volatile.
   C. give them up completely.
   D. are impossible to measure accurately.

6. In naming a binary covalent compound, the IUPAC method orders sulfur before

   A. phophorus.
   B. boron.
   C. carbon.
   D. iodine.

7. Which American chemist first described the energy levels of molecules?

   A. Grover Miller
   B. Linus Pauling
   C. Antoine Lavoisier
   D. Marc Williams

8. What is the electronegativity number of beryllium?

   A. 0.8
   B. 1.5
   C. 1.9
   D. 3.5

9. A dipole moment is the

   A. measurement of the charge separation between each part of a molecule.
   B. distance between tetrahedral angles in a molecule.
   C. Lewis structure containing two sets of double bonds.
   D. unique name for carbon-carbon bonds.

10. If when naming an ionic compound, the first element gets the name root and the second adds on the ending "*ide*," what is the name for LiF?

    A. Cesium fluoride
    B. Lithium fluoride
    C. Iron chloride
    D. Lithium monobromide

chapter 9

Electrochemistry

Electrochemistry is all about electrons experiencing chemical reactions and electricity. Specifically, chemical reactions created by passing an electric current through an ion-containing solution also make it possible for electrons to flow through a wire and generate electricity. In this chapter, the role of electron transfer in electricity will be explained. You will also lean how electrical charge affects chemical reactions along with the importance of oxidation and reduction.

CHAPTER OBJECTIVES

In this chapter, you will

- Understand how oxidation and reduction take place
- Discover the importance of oxidation states
- Figure out how electrolysis is possible
- Learn how electricity is created in voltaic cells

Introduction to Electrochemistry

Since electricity drives certain chemical reactions, chemical and electrical interactions is what electrochemistry is all about. This branch of chemistry is important in photosynthesis, energy storage (batteries), solar cells, electroplating, and the rusting of iron.

Herbert Henry Dow, founder of the Dow Chemical Company, was an electrochemist who tested brine from Michigan wells with trace amounts of bromine. His work in 1889 led Dow Chemical to create electrolysis methods for the extraction of bromine, chlorine, sodium, calcium, and magnesium from ancient seas beneath modern-day Michigan.

What Is Oxidation and Reduction (Redox)?

Oxidation-reduction, or redox, reactions are a group of reactions involving electron transfer. Like pairs' ice skating, it requires two partners. The first partner is oxidation. *Oxidation* is the loss of electrons.

Since the electrons have to go somewhere, oxidation's partner is reduction. *Reduction* is the gain of electrons. Because these processes happen at the same time, oxidation and reduction are called *half-reactions*. The linkage between half-reactions is a lot like acid-base chemistry (Chap. 10), when protons are transferred between compounds.

> **Oxidation-reduction reactions** involve electron transfer between atoms or molecules.

In chemistry, a half-reaction is written as electrons with the atoms. Commonly, the electrons are written as "e⁻" since they are part of the reaction. An example of a half-reaction is as follows:

$$Zn^0 \text{ (s)} \rightarrow Zn^{2+} \text{ (aq)} + 2 \text{ e}^- \text{ (oxidation)}$$

In this half-reaction, solid (s) zinc is undergoing oxidation, and losing electrons. Without those two electrons, the zinc atom would have two more protons than electrons. Remember, electrons are negatively charged, and protons are positively charged. The zinc atom will then have two positive charges and becomes an ion. The (aq) in the equation shows the ion is in solution or "aqueous" in water. Oxidation always makes the net charge of an atom more positive.

EXAMPLE 9-1

A silver ion in solution goes through reduction as follows:

$$Ag^+ (aq) + e^- \rightarrow Ag^0 (s) \text{ (reduction)}$$

The silver ion gains one electron to form solid silver metal. This reaction is balanced. The positive charge of the silver ion plus the electron on one side of the equation makes the total equal to zero. Reduction always makes the net charge on the atom more negative.

? Still Struggling

One way to remember how redox chemistry works is to think of an OIL RIG. The OIL stands for Oxidation Is Loss of electrons. They just slip away. The RIG part is the building up of the atom, because Reduction Is the Gain of electrons.

OIL RIG

Oxidation **I**s electron **L**oss

Reduction **I**s electron **G**ain

Balancing Redox Reactions

When two chemical half-reactions are combined and balanced, the same number and kind of atoms must appear on each side of the half-reaction. The total charge is the same on each side of the half-reaction. Figure 9-1 shows how oxidation numbers increase and decrease during various redox reactions.

Actually, combining half-reactions is just algebra. Take a look at the reaction between silver and zinc. One compound undergoes oxidation, and the other undergoes reduction. This reaction is made up of the combination of the two previous half-reactions:

$$Zn^0 (s) \rightarrow Zn^{2+} (aq) + 2\ e^- \text{ (oxidation)}$$

$$Ag^+ (aq) + e^- \rightarrow Ag^0 (s) \text{ (reduction)}$$

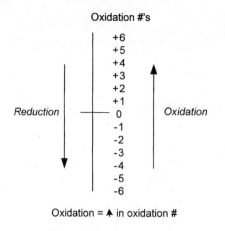

Oxidation #'s

Oxidation = ▲ in oxidation #

Reduction = ▼ in oxidation #

FIGURE 9-1 · Oxidation and reduction are opposite reactions.

However, there is a problem with the electrons. The oxidative half-reaction supplies two electrons, but the reductive half-reaction only needs one. Remember the number of electrons has to be the same for both half-reactions.

The easiest way to solve this problem is to double the silver ions so both zinc electrons are used in the reaction. Doubling the amount of silver ions in the half-reaction doubles the silver metal products. Similarly, other reactions can be combined and balanced by using a multiple.

So, for two half-reactions between zinc and silver, the number of electrons are equal on each side, and can be canceled out as shown:

$$Zn^0 \, (s) \rightarrow Zn^{2+} \, (aq) + 2 \, e^- \, (oxidation)$$

$$2 \, Ag^+ \, (aq) + 2 \, e^- \rightarrow 2 \, Ag^0 \, (s) \, (reduction)$$

$$Zn^0 \, (s) + 2 \, Ag^+ \, (aq) + \cancel{2 \, e^-} \rightarrow Zn^{2+} \, (aq) + 2 \, Ag^0 \, (s) + \cancel{2 \, e^-}$$

$$Zn^0 \, (s) + 2 \, Ag^+ \, (aq) \rightarrow Zn^{2+} \, (aq) + 2 \, Ag^0 \, (s) \, (net \, reaction)$$

Oxidizing Agents

During a redox reaction, an *oxidizing agent* causes the oxidation of another substance. Oxidizing agents are also called oxidants and oxidizers. Common oxidizing agents have electronegative atoms in search of extra electrons. Good examples of electronegative elements are halogens, like fluorine, oxygen, chlorine, and bromine (group VIIA on the Periodic Table).

An **oxidizing agent** causes the oxidation of another substance, while being reduced.

Molecular oxygen (O_2) is an important oxidant in just about everything. For example, when silicon (Si) combines with oxygen and other minerals in the Earth's crust, sand is formed. In fact, most of the compounds found in living organisms on this planet contain oxygen. Human bodies have lots of oxygen since water (H_2O) makes up nearly 60% of the human body by weight.

EXAMPLE 9-2

The examples below show oxygen (O_2) as an oxidizing agent.

$$N_2 + O_2 + (high\ temperature) \rightarrow 2\ NO\ (nitric\ oxide)$$

$$2\ H_2S + 3\ O_2 \rightarrow 2\ H_2O + 2\ SO_2$$

In both reactions, oxygen combines with another substance to make new substances. As the oxidizing agent, oxygen gains electrons from the other substance.

Reducing Agents

During a redox reaction, a *reducing agent* causes the reduction of another substance. Reducing agents are also called reductants and reducers. Remember, a reducing agent gives up electrons to other substances and is oxidized. Common reducing agents have electropositive atoms that give up one or more electrons. Zinc, mercury, tin, magnesium, silver, aluminum, and manganese are all good reducing agents.

A **reducing agent** causes the reduction of another substance, while itself being oxidized.

When hydrogen is heated in combination with metal oxides like copper and zinc, the metal element is separated out and water is formed. The metal oxide is reduced and releases the uncombined metal.

EXAMPLE 9-3

Here hydrogen (H_2) serves as a reducing agent. Notice the change in states (i.e., solid, liquid, gas).

$$CuO(s) + H_2(g) \rightarrow Cu(s) + H_2O(aq)$$

$$ZnO(s) + H_2(g) \rightarrow Zn(s) + H_2O(aq)$$

Reduction also takes place when a compound picks up hydrogen atoms.

EXAMPLE 9-4

Methyl alcohol (CH_3OH) is formed in the reaction of carbon monoxide (CO), hydrogen gas, and a catalyst.

$$CO(g) + 2\ H_2(g) \rightarrow CH_3OH(aq)$$

Oxidation State

Before moving on to reactions, let's look at the oxidation state of an atom. The *oxidation state* makes it a lot easier to figure out what is being oxidized and what is being reduced. Oxidation state shows an atom's degree of oxidation in a chemical compound.

Chemical compounds are made up of different kinds of bonds. If all of an atom's bonds in a compound were completely (100%) ionic, its charge would be its oxidation state. However, not all bonds are ionic and not all atoms are ions.

Knowing an atom's oxidation state is very helpful in solving redox reactions. Figure 9-1 shows the range of oxidation and reduction numbers.

Oxidation states are usually integers that are positive, negative, or zero. In some cases, the oxidation state is an average of possible oxidation states. For example, the oxidation state for iron is +8/3 in magnetite (Fe_3O_4).

When forming oxides with four oxygen atoms, xenon, ruthenium, and osmium have the highest possible oxidation state (+8). The lowest oxidation state (−4) is found in some carbon compounds. Table 9-1 lists a few tips to help in figuring out oxidation numbers.

For example, finding the oxidation numbers for the H_2O molecule is fairly easy. Since it is neutral, all the oxidation states must add up to zero. Oxygen has an oxidation number of −2, so each hydrogen atom must have an oxidation number of +1. That way, the total charge equals zero.

TABLE 9-1 When figuring out oxidation numbers, remembering a few hints can help
General rules of oxidation and reduction
1. An uncombined element has an oxidation number of zero (K, Fe, H_2, O_2).
2. All oxidation numbers added together in a compound must equal zero.
3. In an ion of one atom, the oxidation number is equal to the ion's charge.
4. In an ion of more than one atom, all the oxidation numbers add up to the ion's charge.
5. When oxygen is part of a compound, the oxidation number is 2 [except peroxides $H_2O_2(-1)$].
6. Hydrogen's oxidation number is equal to its +1 charge (except when combined with metals, then it is −1).

Some oxidation states are calculated with a bit more work. If all of the oxidation states in a problem are known except one, you can still find the missing oxidation state. Check out the following example.

EXAMPLE 9-5

What is the oxidation state of lithium dichromate ($Li_2Cr_2O_7$)?

SOLUTION

$$Li = +1, Cr = x, O = -2$$
$$Li_2, \text{ so } 2 \times +1 = +2$$

The anion has two negative charges, $Cr_2O_7^{2-}$

$$Cr_2 = x$$
$$O_7 (7 \times -2 = -14)$$
$$+2 + 2x + (-14) = 0$$
$$2x = 12$$
$$x = +6$$

So chromium's oxidation state is +6.

What Is an Electrochemical Cell?

A very important class of redox reactions is used to create electrical energy with electrochemical cells. A bunch of linked electrochemical cells is called a *battery*.

In 1780, Luigi Galvani, an Italian physician and physicist, made a simple electrochemical cell. When he connected two different metals (e.g., copper and zinc), he could touch each metal to different places in a frog's leg and make it move. He called the observation "animal electricity," because he thought only animals had this talent.

Two decades later, Alessandro Volta, an Italian physicist proved Galvani wrong by creating a *voltaic pile* or primitive battery. Volta made frog-free electricity. Instead, he tried two different metals at opposite edges of a piece of paper wet with salt water.

> A **voltaic pile** is made up of a set of individual galvanic cells placed in series.

In honor of Volta's and Galvani's works, batteries are often called *voltaic* and *galvanic cells*. These cells are also called wet cells, because a solution is used. Not much water is needed. There are also dry cells (e.g., alkaline batteries).

All cells need electrodes to accept and donate electrons during chemical reactions. Oxidation happens at the *anode*. Electrons are pulled away to make ions and mass is lost at this electrode. Reduction takes place at the *cathode*. Electrons are donated to ions, so mass is gained at this electrode. In electrochemistry, anodes and cathodes are used to identify where redox reactions take place.

> **Anode** = electrode where oxidation occurs
>
> **Cathode** = electrode where reduction occurs

? Still Struggling

A simple memory aid for redox reactions is to remember the first letter of each word. *C*athode and *r*eduction both start with consonants. *A*node and *o*xidation start with vowels. Whatever works!

Wet Cells

Although Volta made electricity with chemistry, it took a while before he understood what happened in the wet electrochemical cell. Volta knew something in the liquid was important for the experiment to work. We know now the unknown compounds are *electrolytes*. In the word electrolyte, the *"lyte"* comes from the Greek *"lytos"* meaning able to be dissolved. Sodium, potassium, chloride, calcium, and phosphate are examples of electrolytes.

> **Electrolytes** are any substance or solution with free ions making the substance electrically conductive.

Through experimentation, Volta found silver and zinc were the best metals for electrochemical testing. The reaction in the voltaic pile is as follows:

$$Zn^0 \text{ (s)} + 2\,Ag^+ \text{ (aq)} \rightarrow Zn^{2+} \text{ (aq)} + 2\,Ag^0 \text{ (s) (voltaic pile reaction)}$$

As the reaction takes place, zinc and silver undergo a chemical reaction. Zinc releases electrons to form Zn^{2+} ions at the anode, which absorb into the salt-soaked paper. These electrons pass through the metals to get to the silver side. This movement of electrons is *electricity*.

To show electron movement, a light bulb can be placed between the two metals. As electrons move, the bulb lights up. When electrons get to the other side, they are taken up by silver ions (Ag^+) at the cathode in the salt solution.

The paper has to be between the metals so the solutions with zinc and silver ions are connected. Why? Balance.

The positively charged ions (i.e., cations) must balance their charges. When a silver ion becomes metal, it no longer needs negatively charged ions (i.e., anions) to keep the net charge at zero. The salt solution makes it possible for the anions to exchange partners.

A modern wet cell is shown in Fig. 9-2. The drawing shows a reaction between zinc and copper electrodes placed in two separate baths of sodium nitrate. Oxidation at the zinc electrode (i.e., anode) creates Zn^{2+} ions around the electrode. Reduction at the copper electrode (i.e., cathode) causes Cu^{2+} ions to attach onto the electrode.

The solutions are connected by a salt bridge that acts like Volta's salt-soaked paper. It allows ions to pass across while at the same time balancing ion charges during the reaction. Meanwhile, electrons moving through the wire (from zinc to copper) create electricity.

Another wet electrochemical cell is the lead acid battery. This type has a lead anode and lead dioxide cathode in a solution of sulfuric acid (i.e., electrolyte). Unlike the voltaic pile experiment, the chemical reaction can be reversed with electricity. In a vehicle, this job is done by the engine's alternator.

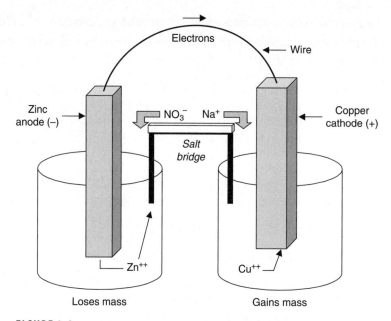

FIGURE 9-2 • Modern wet electrochemical cells are fairly simple.

Dry Cells

Strangely, not a lot of water is needed to make a battery. In 1866, Georges Leclanché, a French engineer, invented the zinc-carbon cell. His idea encompassed a dry electrochemical cell with a conductive electrolyte in a paste instead of a solution.

For the dry cell, a zinc cup is lined with a thin layer of manganese dioxide (MnO_2) and filled with a paste containing the electrolyte ammonium chloride (NH_4Cl) and zinc chloride ($ZnCl_2$). A carbon rod serves as a cathode, while the zinc acts as an anode. The half-reactions include

$$Zn^0 \text{ (s)} \rightarrow Zn^{2+} \text{ (aq)} + 2 \text{ e}^- \text{ (anode)}$$

$$2\,NH_4^+ \text{ (aq)} + 2\,MnO_2 \text{ (s)} + 2 \text{ e}^- \rightarrow Mn_2O_3 \text{ (s)} + 2\,NH_3 \text{ (aq)} + H_2O \text{ (cathode)}$$

$$Zn^0 \text{ (s)} + 2\,NH_4^+ \text{ (aq)} + 2\,MnO_2 \text{ (s)} \rightarrow Zn^{2+} \text{ (aq)} + 2\,NH_3 \text{ (aq)} + Mn_2O_3 \text{ (s)} +$$
$$H_2O \text{ (net reaction)}$$

Simple and inexpensive, the zinc-carbon cell design led to a reliable battery. The zinc-carbon cell was widely used throughout the twentieth century until it was replaced with alkaline cells.

Alkaline cells overcame many of the problems (e.g., short shelf life; decreased power) of zinc-carbon cells. The difference was the replacement of ammonium chloride with potassium hydroxide (KOH). Potassium hydroxide, an alkaline material, caused them to be named alkaline batteries. The half-reactions for an alkaline cell are given as

$$Zn \text{ (s)} + 2 \text{ OH}^- \text{ (aq)} \rightleftharpoons Zn(OH)_2 \text{ (s)} + 2 \text{ e}^- \text{ (anode)}$$

$$2 \text{ MnO}_2 \text{ (s)} + H_2O + 2 \text{ e}^- \rightarrow Mn_2O_3 \text{ (s)} + 2 \text{ OH}^- \text{ (aq) (cathode)}$$

$$Zn \text{ (s)} + 2 \text{ MnO}_2 \text{ (s)} + H_2O \rightarrow Zn(OH)_2 \text{ (s)} + Mn_2O_3 \text{ (s) (net reaction)}$$

What Is Electrolysis?

In 1800, within six weeks of Volta's report, William Nicholson and Anthony Carlisle, two English scientists, discovered *electrolysis* using the voltaic pile. They stuck the ends of the voltaic pile into an electrolytic solution and collected gases from the reaction. Hydrogen formed at the cathode and oxygen formed at the anode.

Through electrolysis, water could be separated into its components. The experimental setup is called an *electrolytic cell*. Electricity is needed for the reaction to occur and is not spontaneous like an electrochemical cell reaction. In fact, an electrolytic cell is the opposite of an electrochemical cell.

> **Electrolysis** is a chemical decomposition reaction created by passing electricity through an ion-containing solution.

Nicholson and Carlisle's experiment was big news. Their results provided solid proof that water was made of hydrogen and oxygen. In fact, there was twice as much hydrogen as oxygen (i.e., H_2O).

They also proved the electricity passing through the system included positive and negative charges with oxygen and hydrogen held together by charged forces. That insight led to modern ideas about chemical bonds. In fact, the discovery of electrolysis started a whole new field of research and applications.

Faraday's Electrolysis Laws

In 1834, Michael Faraday, an English physicist and chemist, published his work on electrochemical reactions and mathematical ways to use electrolysis. Specifically, he described laws connecting mass to electricity. His work led to equations for electroplating metals and producing various chemicals.

First Law of Electrolysis

During electrolysis, the mass of any substance at an electrode (m) is directly proportional to the amount of electricity passed through the solution (q). Mathematically, the first law of electrolysis is written as

$$m \propto q$$

The substance can either be deposited on the electrode or as ions dissolved in solution at the electrode. The amount of electricity needed is found by multiplying the flow of charge (i.e., electrical current) in amperes by time in seconds. The resulting unit (amp/s) is called a *coulomb* (C).

Second Law of Electrolysis

For a specific charge, the mass of a substance at an electrode (m) is directly proportional to its equivalent weight. The equivalent weight is the molar mass (M) divided by the number of electrons transferred per ion (z). The second law of electrolysis is written as

$$m \propto M/z$$

Electrolysis is a really handy process. It's used when separating metal from ores, cleaning archaeological artifacts, and coating materials with extremely thin layers of metal (i.e., electroplating).

Conductors and Insulators

Electricity zips through or is *conducted* through some materials better than others. The *resistance* electricity meets when moving through a substance is measured in *ohm* units. If a material has a low resistance, it conducts electricity better than a material with a high resistance.

The reason for this difference is that some substances, called *insulators*, hold their electrons very tightly. Their electrons can't move easily. Wood, rubber, glass, plastic, and asphalt make good insulators and have very high resistance.

Other materials hold electrons loosely and allow them to speed through easily. These are called *conductors*. Most metals, (e.g., gold, copper, silver, aluminum, or steel) are good conductors.

The SI unit of electrical resistance is the **ohm** (Ω). The opposite measurement, electrical conductance, is measured in **siemens** (S).

For a lot of materials, electrical resistance doesn't depend on the amount of current through or the potential difference (i.e., voltage) across the object. Resistance is constant for a specific temperature and material.

Ohm's law states the current flowing between two ends of a conductor is proportional to the voltage across the two points. The current is inversely proportional to the resistance between them.

A sample's resistance can be defined as the ratio of voltage to current, in accordance with Ohm's law:

$$R = V/I$$

where V is voltage in volts, I is current in amperes, and R is resistance in ohms.

Ohm's law describes the relationships between power (P), voltage (V), current (I), and resistance (R). One ohm equals the resistance through which 1 volt maintains a current of 1 ampere. If you dig deeper into the material sciences and conductivity of metals, you'll learn all about complex electrical properties, but for now, some general terms to know are:

- Current (I)—flows from negative to positive on the surface of a conductor and is measured in (A) amperes or amps
- Voltage (E)—is the difference in electrical potential between two points in a circuit that makes current move and is measured in (V) volts
- Resistance (R)—measures how much current is allowed to flow through a material and is used to control voltage and current levels
- Power (P)—is the quantity of current multiplied by the voltage level at a certain point and is measured in watts

Knowing electrochemistry basics in oxidation-reduction reactions and ionic equilibrium is important. Additionally, electrochemistry and interfacial chemistry (e.g., between two phases like an insoluble solid and a liquid) affect such processes as electrosynthesis, materials corrosion, photoelectrochemical breakdown of pollutants, and the development of electrochemical sensors.

With dwindling resources in many parts of the world, electrochemistry will make materials science applications even more crucial in the future.

QUIZ

1. **An electrochemical cell needs a salt solution to**
   A. provide resistance to the electrodes.
   B. make the solution taste better.
   C. balance the formation and loss of ions.
   D. keep the electrodes wet.

2. **When water is broken down into oxygen and hydrogen gas by an electric current, it is called**
   A. transmutation.
   B. ion deferral.
   C. galvanization.
   D. electrolysis.

3. **The oxidation state for chlorine in $HClO_4$ is**
   A. +7.
   B. −1.
   C. +3.
   D. 0.

4. **Reduction takes place at the *cathode*, and mass is**
   A. unaffected.
   B. lost.
   C. not recorded.
   D. gained.

5. **The electrolysis of water by Nicholson and Carlisle demonstrated that**
   A. water could spontaneously make hydrogen and oxygen.
   B. hydrogen and oxygen are held together by charged forces.
   C. there are two oxygen atoms for every one hydrogen atom.
   D. a salt is not needed for an electrochemical cell.

6. **Mathematically, Faraday's first law of electrolysis is**
   A. $m \propto q$
   B. $m \propto M/z$
   C. $q \propto mc\Delta T$
   D. $E \propto mc^2$

7. **In the reaction $F_2 + H_2 \rightarrow 2\,HF$, fluorine is being**
   A. oxidized.
   B. halogenated.
   C. reduced.
   D. radicalized.

8. **Oxidation state shows an atom's**
   A. atomic weight.
   B. reactant mass.
   C. degree of oxidation in a compound.
   D. ionization temperature.

9. **OIL RIG is a memory aide for**
   A. oxidation is lovely, but reduction is gorgeous.
   B. oxidation is lacking and reduction is loaded.
   C. oxidation is electron gain and reduction is electron loss.
   D. oxidation is electron loss and reduction is electron gain.

10. **What "secret" ingredient made the alkaline cell better than the zinc-carbon cell?**
    A. KOH
    B. NH_4Cl
    C. Zn
    D. MnO_2

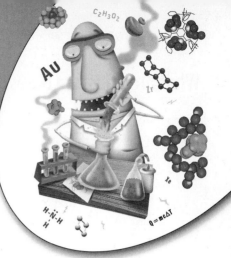

chapter **10**

Acids and Bases

People have known about sour-tasting foods like lemon juice or vinegar for a very long time. In fact, the Greek word for vinegar is *acetum*, from which we get acetic acid, the main acid in vinegar. *Acidus* is the Latin word for sour. But what does acidic and basic really mean? Over the years, the definitions for acids and bases have changed. In this chapter, you will learn all about the pH scale, acids, bases, and neutralization.

CHAPTER OBJECTIVES

In this chapter, you will

- Become familiar with Brønsted-Lowry acids and bases
- Find out why hydrogen is important
- Learn how acids and bases react to become stronger or weaker
- Understand how buffers work
- Learn to handle acids and bases safely

What Are Acids and Bases?

In the seventeenth century, Robert Boyle, best known for Boyle's law, found a way to tell the difference between acids and bases. After testing a lot of different solutions, he created *litmus paper* to test whether a solution was acidic or basic. Litmus paper made from cellulose pulp contains dyes. Acids and bases turn litmus paper different colors. Here is a summary of his findings.

> **Acids** taste sour, corrode metals, change litmus paper to red, and are less acidic when mixed with bases. **Bases** feel slippery, change litmus paper to blue, and are less basic when mixed with acids.

Even though Boyle and others tried to explain why acids and bases behave the way they do, it took almost 200 years before acids and bases were given chemical definitions.

Arrhenius Theory

In the late 1800s, Svante Arrhenius, a Swedish scientist suggested that water dissolves compounds by separating them into distant ions. He thought acids were compounds containing hydrogen. By dissolving an acid in water, hydrogen ions (H^+) are released into the solution. For example, hydrochloric acid (HCl) dissolves in water.

$$\overset{H_2O}{HCl \rightarrow H^+(aq) + Cl^-(aq)}$$

Arrhenius thought bases released hydroxide ions (OH^-) into solution. For example, sodium hydroxide (NaOH) is a typical Arrhenius base.

$$\overset{H_2O}{NaOH \rightarrow Na^+(aq) + OH^-(aq)}$$

The Arrhenius definition of acids and bases explains a lot. Arrhenius's theory clarifies why all acids have similar properties; all acids release H^+ into solution. His theory also explains why all bases have similar properties and release OH^- into solution. While Arrhenius explained how most acids and bases act, he could not explain everything. For example, Arrhenius couldn't explain why some substances (e.g., baking soda ($NaHCO_3$), act like a base even though they don't contain hydroxide ions.

Brønsted-Lowry Acids and Bases

In 1923, a Danish chemist Johannes Brønsted and Thomas Lowry, an English chemist, published separate but related papers on acids and bases. Both men wanted to improve upon Arrhenius's theory about acid and base properties.

Brønsted, for example, discovered that acids and bases could split off or take up hydrogen ions. Their combined ideas became the Brønsted-Lowry definition of acids and bases, which expanded Arrhenius's original idea.

> Brønsted-Lowry acids = proton donors
>
> Brønsted-Lowry bases = proton acceptors

The Brønsted-Lowry definition of acids is a lot like Arrhenius's definition; any substance that can donate a hydrogen ion is an acid. Under Brønsted's definition, acids are known as *proton* donors. A hydrogen ion (i.e., hydrogen minus its electron) is just a proton.

Brønsted bases are different from those Arrhenius described. A Brønsted base is any substance that can accept a hydrogen ion. In other words, a base is the opposite of an acid.

NaOH is still considered a base, because it can accept protons from an acid to form water. (Note: More about this when we look at neutralization.) Specifically, Brønsted-Lowry explained why samples without an OH^- act like bases. Baking soda ($NaHCO_3$), for example, acts like a base by accepting a hydrogen ion from an acid in the following reaction.

$$\text{Acid} \qquad \text{Base} \qquad\qquad \text{Salt}$$
$$HCl + NaHCO_3 \rightarrow H_2CO_3 + NaCl$$

EXAMPLE 10-1

A Brønsted-Lowry reaction is shown. Which is the acid? Which is the base?

$$NH_3 \;+\; H_2O \rightarrow NH_4^+ \;+\; OH^-$$
$$\quad\; base \qquad acid$$

Neutralization

Arrhenius's, Brønsted's, and Lowry's work explained one of Boyle's observations on how acids and bases decreased the strength of the other. This was called *neutralization*. Later scientific advances showed how neutralization took place when protons (H^+) from acids and hydroxide ions (OH^-) from bases combined to make water (H_2O).

$$H^+ (aq) + OH^- (aq) \rightarrow H_2O$$

What else happens during a neutralization reaction? The related ions form a salt, which dissolves in solution.

EXAMPLE 10-2

For instance, when solutions of HCl and NaOH mix, their ions are dissolved by H_2O. The proton and hydroxide combine to form H_2O and a salt from Na^+ and Cl^- ions.

Acid		Base		Water		Salt
HCl	+	NaOH	$\rightarrow$	H_2O	+	NaCl

Neutralization takes place when mixed acids and bases cancel each other out.

Conjugate Acid-Base Pairs

Reactions between acids and bases make *conjugate acids* and *bases*. When an acid reacts and loses a hydrogen ion, it changes into a new compound. With one less proton, the new compound can accept another proton and restore the original acid (i.e., it acts like a base). Because it was made from an acid, it is called a conjugate base.

In the same way, a base can pick up a proton to become a new compound, (i.e., a conjugate acid). Like an acid, a conjugate base donates an extra proton to another molecule. In the previous reaction, H_2O acts like an acid and loses a proton (H^+) to make a conjugate base (OH^-). The base, NH_3, gains a proton to form a conjugate acid NH_4^+.

> A **conjugate base** of a compound or ion forms when a proton is lost, while a **conjugate acid** of a compound or ion forms when a proton is gained.

Conjugate acid-base pairs play a big part in many reactions. Some common acids and bases pick up protons in one reaction, then turn around and donate them in the next reaction. Many biological processes work this way with water or blood acting as the reaction media. Ions or molecules that switch either way are called *amphiprotic*. They can either lose or add a proton (H^+). An ion or molecule is *amphoteric* when it serves as either an acid or a base in a reaction, but has no protons.

Strength of Acids

While all acids give up protons, they are not all alike. Acids that easily give up protons are called strong acids. Sulfuric acid, a strong acid, is the most commonly produced acid in the United States. It is used to make fertilizers (70%) from ores containing phosphate rock, while the rest is used in car batteries, industrial metals, oil refining, and cleaners.

Acids that stubbornly hang on to protons are called weak acids. Boric acid, a weak acid, is used in eye washes, glass cleaners, and insecticides. Many insects can't metabolize boron. Mixing powdered sugar, flour, and borax (i.e., conjugate base of boric acid) creates a bait that kills cockroaches. The boron molecule clogs the insects' digestion so they die, but it is safe for people and animals. Table 10-1 lists some common acids from strongest to weakest.

TABLE 10-1	Common acids from strongest to weakest should be handled with matching caution
Name	**Formula**
Hydrochloric acid	HCl
Sulfuric acid	H_2SO_4
Nitric acid	HNO_3
Perchloric acid	$HClO_4$
Phosphoric acid	H_3PO_4
Acetic acid	CH_3COOH
Citric acid	$C_3H_5(COOH)_3$
Lactic acid	$CH_3CHOHCOOH$
Boric acid	H_3BO_3

Acid strength depends on how well it gives up protons or how well it ionizes. (Note: See section on ions in Chap. 8.) As discussed in the previous section, this process generates a conjugate base.

A conjugate's strength is opposite to that of the original acid. Strong acids are weak conjugate bases. When a conjugate base is protonated, the strong acid changes and throws off a proton. The reaction is back to where it started. On the other hand, weak acids make strong conjugate bases. A weak acid doesn't give up its protons easily, so its conjugate base keeps protons to form the original acid. The reaction favors the weaker compound, whether an acid or conjugate base.

> Strong acids = weak conjugate bases
>
> Weak acids = strong conjugate bases

There are clues to figuring out acid strength. First, bond polarity with a hydrogen atom is important. The more polar the bond, the easier the proton is pulled away. Second, the size of the atom bonded to the hydrogen is also important.

Usually, larger atoms form weaker bonds so a proton is more willingly given away (i.e., bigger atoms make stronger acids). When comparing the acids HF, HCl, HBr, and HI, acidity increases as the atom's size increases. In fact, acid strength can be arranged in increasing order:

$$H_2O < HF < HCl < HBr < HI$$

Strength of Bases

While all bases accept protons, they do not accept them equally. Bases, that take up protons easily, are called strong bases. A common laboratory base, sodium hydroxide (NaOH), is used by many chemistry students. The metal sodium (Na^+) forms a tight bond with the (OH^-) ion until dissolved in H_2O. Sodium hydroxide is a strong cleaner, soil stabilizer in road construction, and soap ingredient.

Some bases don't take up protons easily and are known as weak bases. Calcium carbonate ($CaCO_3$), found in antacids, is a weak base. When spicy or acidic foods like tomatoes cause heartburn, an antacid containing calcium carbonate neutralizes stomach acids and brings relief. Table 10-2 lists a few commonly used bases.

TABLE 10-2	It is just as important to become familiar with chemical bases as with acids

Name	Formula
Sodium hydroxide	NaOH
Potassium hydroxide	KOH
Calcium hydroxide	CaOH
Magnesium hydroxide	$Mg(OH)_2$
Calcium carbonate	$CaCO_3$
Ammonia	NH_3
Calcium diphosphate	$Ca(H_2PO_4)_2$
Barium sulfate	$BaSO_4$

Base strength depends on how easily protons are accepted. As mentioned earlier, this process generates a *conjugate acid*. The conjugate's strength is opposite to that of the original base. Strong bases make weak conjugate acids. When the conjugate acid loses a proton, a strong base is re-formed. The strong base then picks up another proton. The reaction has boomeranged back to where it began.

Weak bases make strong conjugate acids. A weak base doesn't accept protons well, so its conjugate acid releases protons to form the original base. The reaction ultimately favors the weaker compound whether base or conjugate acid.

> Strong bases = weak conjugate acids
> Weak bases = strong conjugate acids

Like acids, there are clues to base strength. Stronger bases ionize in water almost totally, while weak bases don't. This happens because less ionized bases have no free orbitals to accept more protons. In other words, everybody is comfortable, stable, and resistant to change. A stronger base is able to accept protons more easily than a weaker one, because it becomes an ion more easily. The following bases are arranged by strength $OH^- > NH_3 > HCO_3^- > CH_3CO_2^- > NO_3^- > HSO_4^-$. Table 10-3 shows the various strengths and weaknesses of acids and bases.

TABLE 10-3 Many foods and household solutions are strongly acidic or basic

Name	pH
Hydrochloric acid	2.0
Stomach acid	1.0–3.0
Lemon juice	2.2–2.4
Vinegar	2.4–3.4
Carbonated drinks	3.9
Beer	4.0–4.5
Milk	6.4
Blood	7.4
Sea water	7.0–8.3
Baking soda	8.4
Antacid	10.5
Cleaning ammonia	11.9
Bleach (sodium hydroxide)	14.0

Why Is Hydrogen Important?

Early researchers eventually figured out that water was not a single element but made of separate elements (i.e., hydrogen and oxygen). In fact, hydrogen gets its name from the Greek words *hudor* or water, and *genna*, meaning to generate.

Hydrogen was described in 1766 by Henry Cavendish. Made of one proton and one electron, hydrogen has an atomic number of 1 and is the first element of the Periodic Table. A nonmetallic colorless, tasteless, odorless gas at 298 K, hydrogen is highly flammable in the presence of oxygen and reacts with most elements.

Stars have an enormous supply of hydrogen. In fact, hydrogen is the most abundant element in the universe, making up over 90% of the visible universe's mass. On Earth, hydrogen combines with oxygen to form water which covers over 70% of the Earth's surface. However, in the Earth's crust, hydrogen makes up only about 0.9% of the composition.

Hydrogen Compounds

Hydrogen reacts with most elements, especially carbon, when it combines and forms starches, hydrocarbons, fats, oils, proteins, and enzymes. We will learn more about hydrogen and carbon bonding in Chap. 12.

Hydrogen and nitrogen form ammonia and salt-forming compounds or *halogens* (e.g., chlorine, fluorine, bromine, iodine, and astatine) in acidic hydrogen halides. It combines with sulfur in the rotten egg smell (hydrogen sulfide, H_2S), and with oxygen and sulfur in sulfuric acid (H_2SO_4).

Hydrogen is also important in the hydrogenation of oils, methanol production, rocket fuel, welding, production of hydrochloric acid, and reduction of metallic ores. It is an important part of cryogenics methods and superconductivity experiments since its melting point is just above absolute zero.

Hydrogen got a bad reputation in 1937 when engineers used it as the lifting gas for the huge airship, *Hindenberg*. Later, it became clear that the Hindenberg's tragic crash and burn was not caused by its hydrogen, but by static electricity setting fire to the varnish on the airship's fabric covering, igniting the hydrogen within and causing the disaster. Modern airships or *blimps*, use an nonreactive gas like helium for lift.

Hydrogen Ion

During reactions, a hydrogen ion is called a proton since it carries a single positive charge. In liquids, the H^+ ion is fully bonded or *hydrated* [e.g., hydroxonium ion (H_3O^+)].

In the early 1800s, while working out the details of the atomic theory, John Dalton burned hydrogen gas in oxygen to see whether or not an atom had a particular mass. His results showed 1 g of hydrogen reacts with 8 g of oxygen. It wasn't until much later that chemists figured out hydrogen actually bonds with two atoms of oxygen (i.e., atomic weight of roughly 16 grams) to form water.

pH Scale

While Boyle's litmus paper worked great for finding if a solution was acidic or basic, it couldn't give an idea about its strength. In 1909, Søren Sørensen, came up with the *pH scale* to measure acidity. The name of the pH scale, from the French for "power of hydrogen," measures hydrogen ions present in a solution.

> **pH scale** measures acidity of a solution based on hydrogen ion (proton) concentration.

In the Brønsted-Lowry definition, acids and bases are linked to the concentration of hydrogen ions. Acids raise the hydrogen ion concentration, while bases lower it because they accept hydrogen ions. The acidity or basicity of a solution, then, is measured by its hydrogen ion concentration. Sørensen's pH scale is shown in the following formula:

$$pH = -\log [H^+]$$

where $[H^+]$ hydrogen ion concentration.

In measuring pH, $[H^+]$ is in moles of H^+/liter solution. For example, a solution with $[H^+] = 1 \times 10^{-7}$ moles/liter has a pH equal to 7.

? Still Struggling

To find pH, just remember to check the exponent. It is equal to pH without the minus sign. The $[H^+]$ = hydrogen ion concentration for vinegar is 1×10^{-3}. The pH of vinegar is 3.

pH of a base > 7

pH of an acid < 7

The pH scale ranges from 0 to 14. Acids have a pH between 0 and less than 7. Acidity and pH are inversely related. Lower pH means higher $[H^+]$ and a stronger acid. Bases have a pH greater than 7 and up to 14. Oddly enough, basicity and pH are directly related. In other words, high pH means lower $[H^+]$ and a stronger base.

Neutral substances, like pure water, have a pH 7. Under those conditions, H^+ and OH^- are in equal numbers. The relationship between $[H^+]$ and pH is shown in Table 10-3 along with acids and bases in everyday life.

pH Meter

Probably the piece of equipment most used in any laboratory beside the beaker or scale is the pH meter. A pH meter has a special sensor which tests liquid samples for acidity or basicity. By sending a current through the sample and measuring resistance/current changes, the amount of hydrogen ions (positively charged ions) can be found. Figure 10-1 shows a typical pH meter.

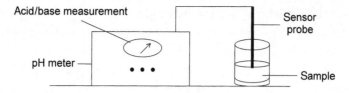

FIGURE 10-1 · A standard pH meter uses a probe to test the acidity of a sample.

Titration

Titration is another way to find an unknown solution's concentration. For example, a basic solution can be added slowly (e.g., drop by drop) to an acidic solution of unknown concentration until it changes color. When this happens, you know you've reached the *end point* and no more base is needed.

> **Titration** is a laboratory method, which makes it possible to find the concentration of a particular acid/base through neutralization.

❓ Still Struggling

The whole idea of titration is to find the equivalence point. The *equivalence point* is the point where the number of moles of a base equals the number of moles of an acid. If the right color indicator is chosen, the end point nearly equals the equivalence point. This is because a solution's pH and color changes very quickly close to the equivalence point because of the steepness of the pH change.

Buffers

Buffers bring many previously discussed concepts together. Buffers can slow or block pH changes. Since pH affects chemical reactions, it should be controlled for an experiment to work. Buffers offer flexibility in adjusting a solution's pH.

> **Buffers** bind and release protons to lessen changes in pH.

For this to happen, a buffering compound must be able to accept or release protons while blocking big pH swings. There must be equal concentrations of an acid and the conjugate base or a base and the conjugate acid. These molecules grab or release protons to minimize pH changes. Not all buffers are created equal. Each buffer excels at a particular pH depending on the strength of the acid or base (and its conjugate).

There is a limit to what buffers can do. This is known as its *buffering capacity*. If too many protons or hydroxide ions are added, the buffer becomes overpowered and pH shifts high or low as a result.

? Still Struggling

Think of buffering capacity like a trip to the market. Your first few packages are easy to hold. As you continue to buy more and more, you must shuffle and reshuffle the bags, boxes, and envelopes in your hands. You are able to maintain balance and hold all your purchases, up to a certain point, but when you go past that, the whole juggling act tips and everything falls to the ground.

Acids, Bases, and Safety

It's important to remember that acids have very low pH and can burn through skin, clothes, shoes, and almost everything they touch. They don't burn with a flame as a fire does, but by reacting strongly with the atoms of a substance, permanently damaging them. Acids and bases *both* cause chemical burns.

As anyone who has ever gotten a strong acid on the skin can tell you, the reaction of acid with skin molecules is bad news. But what if an accident happens and you or your lab partner are splashed with acid? How would you stop the reaction?

Get under a water spray fast! Whether washing your hands, using an eye wash, or being doused in a full-body shower, water dilutes acids. A base will counteract the acid, but this usually takes too much time.

Water is the best answer for strong base spills, too. Since they are fully ionized, strong bases cause the same kinds of intense reactions as acids. Sometimes students forget this and treat bases with less respect than acids. Don't do it! Strong bases cause the same kinds of bad burns as strong acids.

Students are warned to be careful when handling strong acids and strong bases in their first laboratory class. Here are some important lab safety tips including

- Always remember to *handle strong acids and bases very carefully*.
- Don't eat or drink anything in the lab.
- Keep fingers out of the burner flame.
- Don't blow yourself up!

Hydrochloric acid (HCl) is one of the first acids a beginning chemistry student encounters. This strong acid reacts with most metals and forms hydrogen ions. Many experiments use HCl as a reactant. It's easy to spot a sloppy chemistry student, just look at his or her clothes and shoes. If there are lots of tiny holes in them, chances are good they have splashed an acid or base while pouring. This is why chemists use laboratory coats and eye protection.

You've learned how acid and base chemistry focuses on a compound or solutions' reactivity in the presence of strong and weak acids and bases. With this skill, you'll have a much better handle on understanding chemistry overall.

QUIZ

1. **Bases**
   A. have a pH <7.0.
   B. do not cause chemical burns.
   C. cannot be used to neutralize acids.
   D. do not dissolve in water.

2. **The pH scale measures a liquid's**
   A. weight.
   B. temperature.
   C. acidity.
   D. density.

3. **All of the following are common "rule of thumb" ways to determine whether or not a solution is an acid, except**
   A. in dilute solutions, acids taste bitter.
   B. litmus paper changes from blue to red.
   C. acids react with metals like iron, magnesium, and zinc and release hydrogen gas.
   D. when combined with bases, the products are water and salt.

4. **The primary acid found in common vinegar is**
   A. boric acid.
   B. hydrochloric acid.
   C. acetic acid.
   D. sulfuric acid.

5. **A strong acid has the following characteristics, except**
   A. combined with a base it gives off a very pungent odor.
   B. it has a pH value >8.5.
   C. it completely ionizes in water and gives up a proton to water to form a hydronium ion, H^3O^+.
   D. it has a strong conjugate base.

6. **A buffer usually contains either a weak acid or base and it's associated with a**
   A. heavy metal.
   B. gas.
   C. salt.
   D. crystal structure.

7. When an ion or molecule can serve as either an acid or base in a reaction, but has no protons (H⁺) it is said to be
   A. ambidextrous.
   B. amphotonic.
   C. amphiprotic.
   D. amphoteric.

8. A base is any solution that releases hydroxide (OH⁻) ions in water and has a pH of
   A. < 7.0.
   B. > 7.0.
   C. = 7.0.
   D. zero.

9. Acids and bases both
   A. cause chemical burns.
   B. react with ideal gases.
   C. taste bitter.
   D. change litmus paper from red to blue.

10. All of the following are properties of Brønsted-Lowry acids and bases, except
   A. an acid provides protons in a reaction.
   B. some reactants can swing either way, providing or accepting protons.
   C. a base accepts protons.
   D. acids have the same proton affinities in every reaction.

chapter 11

Thermodynamics

In this chapter, you will learn how energy is converted to heat and mechanical work, as well as the differences between enthalpy and entropy.

CHAPTER OBJECTIVES

In this chapter, you will

- Understand the first, second, and third laws of thermodynamics
- Find out how enthalpy affects reactions
- Learn the difference between potential and kinetic energy
- Discover the importance of activation energy
- Understand how reaction order relates to the reactants' concentrations

What Is Thermodynamics?

The word *thermodynamics* comes from the Greek words for heat (*therme*) and power (*dynamis*). Thermodynamics covers the conversion of energy to heat and mechanical work.

Think about roasting marshmallows around a campfire. Energy within the wood is transformed by the fire and released as heat, a form of energy. A marshmallow on a stick is made of sugar and other ingredients that absorb heat causing it to melt. When the marshmallow gets too hot, it catches fire and burns, releasing more heat. In this simple activity, there's a whole lot of thermodynamics going on.

Potential and Kinetic Energy

Thermodynamics focuses on the idea that there is a certain amount of *internal energy* (*U*) in an isolated system. This energy is made up of the *potential* and *kinetic* energy of a system's atoms and molecules conveyed as heat.

Potential energy is stored within a system. If a rubber band is stretched, it contains potential energy. Kinetic energy is the energy of motion. If a stretched rubber band is released, it will fly away with kinetic energy. The potential energy in the rubber changes into kinetic energy.

The amount of kinetic energy a moving object has depends on its mass and how fast it is moving. The greater the object's mass, the greater its speed and the more kinetic energy it has.

In the following equation, kinetic energy relates mass (*m*) and velocity (*v*):

$$\text{Kinetic energy} = 1/2 \times m \times v^2$$

Kinetic energy describes the energy an object has due to *motion*.

What Is Standard State?

Thermodynamics quantities are often listed for *standard state* conditions, so experimental results can be easily compared. Standard temperature and pressure are often shortened to STP and equal to 25°C and 1 atmosphere (atm). When scientists look at a system's standard state, they examine pressure, temperature, moles, and physical state (e.g., liquid, crystal, gas) conditions. An element's standard or *reference state* is defined as its most stable form at 1 bar pressure and a set temperature (e.g., 298.15 K).

The **standard state** of a pure substance, mixture, or solution is the baseline for finding different characteristics under changing conditions.

State Functions

State functions describe a system's present state, but not how it got that way. A state function gives a system's current equilibrium state. Gibbs free energy, entropy, enthalpy, and energy are all state functions. They describe a thermodynamic system's equilibrium state. By contrast, work and heat are not state functions. They describe changes in a system's dynamic equilibrium. You'll learn more about these later in the chapter.

First Law of Thermodynamics

The *First Law of Thermodynamics*, also called the *law of conservation of matter*, describes how energy can be changed from one form to another, but without being created or destroyed.

In practice, this law relates potential energy (U), heat (Q), and work (W). Any changes in potential energy involve heat and work. Specifically, heat energy is added to a system from its surroundings. But the total energy can't be lost or created.

A thermodynamics system loses energy in the form of work, which is given back to the surroundings (i.e., total gains and losses equals zero). Because these variables are changing, the Greek letter *delta* (Δ) is added to each to show a change. Mathematically, this relationship is written as

$$\Delta U = \Delta Q - \Delta W$$

? Still Struggling

An engine's piston illustrates this idea. If the gas in a piston is heated, then energy is introduced into the system. The gas expands and increases its volume, which in turn pushes up on the piston. This causes the system to generate energy as work.

Work is usually done by a gas. If the pressure is constant, then work is equal to the pressure (P) multiplied by the difference in volume (ΔV) from an initial state to a final state. The first law of thermodynamics can then be written as:

$$\Delta U = \Delta Q - P \times \Delta V$$

$$\Delta V = V_{final} - V_{initial}$$

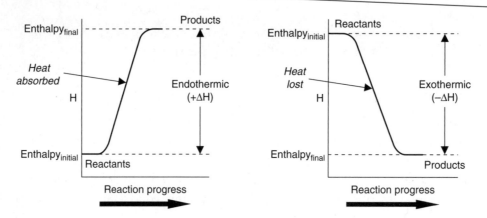

FIGURE 11-1 • Heat is absorbed or lost when reactants are changed into products.

Enthalpy

One way of representing Q is by *enthalpy* (H). Enthalpy is the amount of heat in a sample. A system's total enthalpy cannot be measured directly, but it is possible to measure a change in enthalpy, ΔH.

For example, a change in enthalpy may be the heat given off or absorbed during a chemical reaction. In an endothermic reaction (i.e., heat absorbed), ΔH is positive, while an exothermic reaction has a negative ΔH since heat is generated. Figure 11-1 shows the process of these types of reactions. Enthalpy units are as follows:

$$kJ \cdot mol^{-1} \text{ (kJ/mol) or } kcal \cdot mol^{-1} \text{ (kcal/mol)}$$
$$1 \; calorie \; (1 \text{ cal}) = 4.184 \text{ joules } (4.184 \text{ J})$$

For gases, enthalpy (H) is equals the sum of internal energy (U) and the product of the pressure (P) and volume (V).

$$H_{sys} = U + P \times V$$

Standard Heat of Formation

In a compound, the standard heat of formation (ΔH_f°) equals the change in enthalpy for the formation of 1 mole of sample (i.e., at standard state) from the initial elements in their standard states, such as 1 atm and 25°C. The standard heat of formation is measured in units of energy per amount of sample. Most are defined in kilojoules per mole as well as kilocalories per mole.

For example, the standard heat of formation of carbon dioxide is described by the following reaction under standard conditions:

$$C \text{ (graphite)} + O_2 \text{ (g)} \rightarrow CO \text{ (g)}$$

Heat of Reaction

Heat of reaction is gained by or removed from a reaction so temperature doesn't change. If reaction pressure in a container is kept constant, then the heat of reaction corresponds to the enthalpy change or heat content. If a sample is dissolving in solution, then the heat of reaction is actually a heat of solution.

Exothermic ($\Delta H < 0$) = Heat ⬆ (lost), temperature ⬆

Endothermic ($\Delta H > 0$) = Heat ⬇ (absorbed), temperature ⬇

Hess's Law

In *Hess's law* energy change depends only on the reactant and product states not on the steps between the states. Energy gained or lost during a chemical reaction is the same whether the reaction happened in one step or many. The total energy change equals the sum of the energy changes leading to the overall reaction. In other words, the enthalpy of a reaction is the difference between the enthalpy of the products minus the enthalpy of the reactants. In mathematical terms, it is written in the following way:

$$\Delta H_{reaction} = \Delta H_{products} - \Delta H_{reactants}$$

Heat of Vaporization

When keeping temperature constant, the total heat needed to convert a hot liquid into a vapor is called the *heat of vaporization* (ΔH_{vap}) or *heat of evaporation*. Measured at a sample's boiling point and atmospheric pressure (1 atm), it is often corrected to 298 K. Water's heat of vaporization is 540 cal/g at 100°C. Due to energy conservation, the condensation of gas to liquid releases heat.

Heat of Fusion

Without changing temperature, the amount of heat absorbed when a solid sample changes into a liquid at its melting point is called the *heat of fusion* or *heat of solidification*. Water's heat of fusion is 80 cal/g at 0°C. Again, due to the conservation of energy, liquids release the same amount of heat when they solidify. A sample's heat of fusion is one of its fundamental properties.

Calorimetry

Joseph Black, a Scottish physician, is thought to be the founder of *calorimetry*. The word comes from *calor*, which is Latin for heat. Heat can be either gained or lost from a system through exothermic or endothermic reactions. Even changes between solid, liquid, and gas phases have heat transfers that can be measured.

Calorimetry is the measurement of heat created or absorbed in a chemical reaction, change of state, or formation of a solution.

The heat measured by calorimetry experiments depends on the sample mass (*m*), *heat capacity* (*c*), and temperature change. *Heat capacity* is the amount of energy needed to increase the temperature of a sample 1°C. This function depends on the chemical properties of each substance. *Molar heat capacity* describes the heat needed to raise the temperature of *1 mole* of a sample by 1°C.

Similarly, *specific heat* is the heat energy needed to raise the temperature of 1 gram of a sample by 1°C. The specific heat per gram of water is much higher than for a metal. In practice, calorimetry uses the following equation:

$$q = m \times c \times (T_{final} - T_{initial})$$

Heat capacity depends on a substance's specific chemical properties. The heat capacity of water is higher than any other common substance. While the heat capacity of ethanol is 2.4 J/g·K, the heat capacity of liquid water is 4.187 J/g·K. As a result, water is important in the weather's temperature changes. The high amount of heat needed to change water's temperature regulates any big shifts in temperature. This is why misting fans feel so cool on a hot day.

EXAMPLE 11-1

During a class chemistry experiment, 1.8 g of sodium hydroxide [NaOH$_{(s)}$] pellets is dissolved in 100 mL of water at 25°C. If the water temperature goes up by 2.5°C, what is the ΔH (*heat of solution*) for the reaction?

✔ SOLUTION

- Find the amount of heat generated, in joules:

$$q = m \times c \times (T_{final} - T_{initial})$$
$$q = 100 \times 4.184 \times (27.5 - 25) = 1046 \text{ J}$$

- Find the moles of solute (NaOH):

$$n = m/\text{molecular weight}$$
$$n \text{ (NaOH)} = 1.8/(22.99 + 16.00 + 1.008) = 0.045 \text{ mol}$$

- Find the change in enthalpy, ΔH, for kilojoules per moles of solute, which is the ratio of heat divided by the number of moles in the experiment:

$$\Delta H = [-q/1000]/n \text{ (solute)}$$
$$[-1046/1000]/0.045 = -23.24 \text{ kJ/mol}$$

(Note: 1000 was included to convert heat in joules to kilojoules.) Because the reaction is exothermic, ΔH is negative.

Second Law of Thermodynamics

The *second law of thermodynamics* explains that the entropy of an isolated system always increases, and is often called the *law of increased entropy*. Entropy (S) is disorder or chaos in a system. In other words, entropy measures the energy in a system unavailable for work. The increase in lost work energy is directly related to an increase in a system's disorganization.

In mathematical terms, $S > 0$. This is the natural disorder of matter and energy, even though the amount stays the same (i.e., law of conservation of matter). The usable energy is permanently lost in the form of unusable energy.

Entropy is the measure of unavailable energy within a closed system (e.g., the universe).

? **Still Struggling**

If a glass of hot water is placed in a cold room, the heat will automatically trans-fer from the glass to the room. This takes place until the water's temperature and the room's temperature are equal. The temperature difference drops and the system's entropy goes up. The second law works in the same way. Over time, differences in temperature, pressure, and chemical potential balance out in an isolated system. Entropy measures how much balancing has taken place.

Third Law of Thermodynamics

The *third law of thermodynamics* is the least common of the three laws. It explains that as a temperature approaches *absolute zero* on the kelvin scale, the entropy of the system reaches a minimum, but not zero ($S > 0$).

> **Absolute** zero is equal to zero on the kelvin temperature scale, which is −273.15°C or −459.7°F.

However, because of the second law of thermodynamics no substance or system can have a temperature of 0 K. The second law encompasses the idea that matter can't move from a colder state to a hotter state instantly, but must go through several energy changes. So, as a sample nears absolute zero, it must pull energy from its environment and can't reach absolute zero.

Gibbs Free Energy

In the late nineteenth century, an American physicist, Josiah Gibbs developed the theoretical basis for chemical thermodynamics and physical chemistry. He suggested a thermodynamic unit, which combined enthalpy and entropy into a single value called *free energy* (G).

Gibbs free energy is an amount of potential work available from an isother-mal, isobaric thermodynamic system. (Note: *iso* is a prefix meaning same or unchanged.)

In an *isothermal process*, a system is kept at the same temperature during an experiment, such that $T_{initial} = T_{final}$. An *adiabatic process* happens when the system is thermally isolated so that no heat can enter or exit. An *isobaric process* is a thermodynamic method where pressure doesn't change, or $P_{initial} = P_{final}$.

Under these conditions, Gibbs free energy is a system's ability to do nonmechanical work or have nonmechanical work done to it. Mathematically, free energy of a system (G) is the sum of its enthalpy (H) and the product of the temperature (T in kelvin) and entropy (S) of the system:

$$G(p,T) = U + pV - TS$$
$$G = H - T \times S$$

During chemical reactions, a system changes from an initial state to a final state. The Gibbs free energy equation for a reaction is then written with a (Δ) indicating change.

$$\Delta G = \Delta H - T \times \Delta S$$

Because the system is closed, the Gibbs free energy for the reaction is the difference in free energy between the products and reactants of the reaction.

$$\Delta G_{reaction} = \Delta G_{f\,products} - \Delta G_{f\,reactants}$$

Like the heat of formation, there is a standard free energy of formation for a sample under standard conditions like 1 atm and 25°C. The standard free energy of formation is written as ΔG_f. The value for an element is zero.

When calculating ΔG, remember to be careful about the units. Entropy is commonly written in joules per kilogram, while enthalpy is usually shown as kilojoules per mole.

Chemical Kinetics

Both thermodynamics and *chemical kinetics* involve reactions. This is seen in ion exchange, acid-base interactions (Chap. 10), and the making or breaking of covalent bonds (Chap. 8). However, chemical kinetics is all about speed. Kinetics comes from the Greek word *kinesis*, which means motion. Many different units describe reaction rates, such as mole per second, grams per second, pounds per second, and kilograms per day, or even changes in concentration like mole per liters-second and gram per liters-second.

Chemical kinetics is the study of reaction rates and mechanisms.

? Still Struggling

Chemical kinetics is centered around experimental conditions (e.g., total reaction speed and reaction mechanisms). For example, temperature and sample concentration affect reaction rates. Mathematical models help explain reaction behavior under different conditions. Sometimes reaction rate involves contributions from many rates and at other times just a single reaction step.

When a single step is critical to the reaction, it is considered *rate-limiting*, because the reaction doesn't happen until the rate-limiting step takes place. Let's take a look at reaction rates in the next section.

Activation Energy

In 1889, Svante Arrhenius, a Swedish scientist, introduced *activation energy* (E_a). It's an energetic blockage between reactants and products that must be cleared before a reaction can occur. Activation energy, then, impacts chemical reaction rates. The greater the energy needed for the reaction to occur, the slower the reaction rate.

> **Activation energy** describes the lowest amount of energy needed for a reaction to take place.

There are big differences in the activation energy for various reactions. Some elements and compounds interact just by being around each other and don't need any added energy (e.g., heat, light, or electricity). Their activation energy is zero.

Other elements react only after a certain amount of energy is added. Lighting a match on a matchbox, for example, provides the activation energy (e.g., heat from friction) needed for the match head chemicals to catch fire. Activation energy is most often written in joules per mole of reactants (J/mol). Figure 11-2 illustrates the activation energy needed for a reaction to take place between reactants.

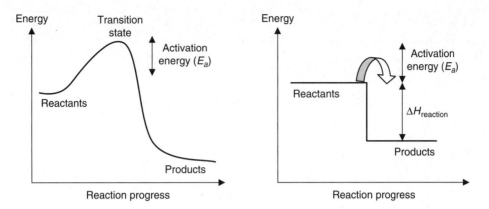

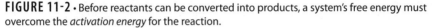

FIGURE 11-2 · Before reactants can be converted into products, a system's free energy must overcome the *activation energy* for the reaction.

Catalysts lower a reaction's activation energy, while not being used up during the reaction. By clearing reaction blockages, catalysts speed reaction rates, without affecting the reaction rate equilibrium. Catalysts aren't thought of as reactants, since they aren't used up during the reaction, but are recycled at the end of a reaction to start another reaction.

> A **catalyst** decreases the activation energy needed to start a reaction process, but doesn't get used up in the process.

Activation Energy and Temperature

The higher the temperature, the faster a reaction happens. To remember this concept, think about walking across hot coals. The extreme heat speeds your reaction and super heated feet considerably.

Systems have more kinetic energy at higher temperatures, so molecules are more likely to collide. With more collisions, it is more likely the right impact will take place and help a reaction overcome its activation energy.

> **Arrhenius rate equation** links a chemical reaction's rate constant to the exponential value of temperature.

Arrhenius developed the mathematics (i.e., connecting temperature and activation energy) that explains and predicts chemical reaction results. The equation is shown below:

$$\ln(k) = E_a/R\,[1/T] + \ln(A)$$

where k = rate constant
E_a = activation energy
R = gas constant (i.e., 8.31 J/k·mol)
T = absolute temperature (kelvin)
A = constant related to collision frequency and configuration

Reaction Order

A reaction order describes the connection between reactants, their concentrations, and an observed reaction rate. The reaction order for a certain reactant is the power to which the concentration is raised in the rate equation. Reaction order is found through experimentation. It can't be figured out from reaction coefficients. Table 11-1 shows the different types of equations and their reaction orders.

For example, a rate equation is shown for the following reaction:

$$2A + B \rightarrow C, \text{Rate} = k \times [A]^2 \times [B]^1$$

where k = a constant for the particular reaction. Reaction order is $A = 2$ and $B = 1$. The total reaction order equals the sum of $A + B$, that is $C = 3$. While these numbers work out simply, reaction order is not always a whole number; it can be zero or even a fraction. It depends on the reaction.

TABLE 11-1 Equations and reaction orders

Type of Reaction	Rate Equation	Order of Reaction
$A \rightarrow B$	Reaction rate = $k\,[A]$	First–order reaction (impact needed)
$A + B \rightarrow C + D$	Reaction rate = $k\,[A][B]$	Second–order reaction
$A + A \rightarrow D$	Reaction rate = $k\,[A]^2$	Second–order reaction

Still Struggling

Take a look at the hydrolysis of ethyl acetate. The normal reaction and rate equation for this reaction are:

$$CH_3COOC_2H_5 \text{ (ethyl acetate)} + OH^- \rightarrow CH_3COO^- \text{ (acetate)} + C_2H_5OH \text{ (ethanol)}$$

$$\text{Rate} = k \times [CH_3COOC_2H_5] \times [OH^-]$$

If imidazole is added to the reaction, the rate equation changes. Imidazole, the catalyst isn't used up. Because it's not a reactant, imidazole isn't written into the reaction. However, it is shown in the rate equation below since catalysts still have a role in reactions.

$$\text{Rate} = k \times [CH_3COOC_2H_5] \times [\text{imidazole}]$$

EXAMPLE **11-2**

See the first-, second-, and third-order reaction equations below. In this first-order reaction, the exponent of a single reactant's concentration equals 1:

$$\text{Rate} = k \times [A]^1$$

In this second-order reaction, the exponent of $B = 2$.

$$\text{Rate} = k \times [A]^2 \times [B]^2$$

In this third-order reaction, the exponents of all the reactants in the rate law add up to 3.

$$\text{Rate} = k \times [A]^2 \times [B]^1$$

EXAMPLE **11-3**

Take a look at the reaction: $2H_2 + 2NO \rightarrow N_2 + 2H_2O$
The rate equation is $r_A = k \times [H_2] \times [NO]^2 - H_2$ (Note: No superscript is first order; NO is second order.)
The rate equation shows three reaction steps:

$$2\,NO \leftrightarrow N_2O_2 \text{ (speedy equilibrium, } k\text{)}$$

$$N_2O_2 + H_2 \rightarrow N_2O + H_2O \text{ (slow, } k_2\text{)}$$
$$N_2O + H_2 \rightarrow N_2 + H_2 \text{ (fast, } k_3\text{)}$$

Of these, the slower middle reaction is the *rate-limiting* or rate-determining step.

Equilibrium

Law of Mass Action

Around the middle of the nineteenth century, Cato Maximillian Guldberg and Peter Waage came up with the math to predict how solutions behave in equilibrium, the *law of mass action*. This law describes a system at equilibrium with an *equilibrium constant* (K_{eq}). The law also links rate constants to a system's equilibrium. The equilibrium constant is the ratio of the forward rate divided by the reverse reaction rate.

$$K_{eq} = k_{\text{forward reaction}} / k_{\text{reverse reaction}}$$

> The **law of mass action** explains how a specific chemical's reaction rate is proportional to the sum of its reactant concentrations.

Equilibrium Constant

The equilibrium constant illustrates how strongly an equilibrium favors one end of a reaction over another. As reactants create products, the greater the equilibrium constant, the more products are formed in the reaction. There are two ways to calculate a reaction's equilibrium constant. The first is to divide the forward reaction rate by the reverse rate. The second uses the reactants' and products' concentration at equilibrium.

For example, if reactants A and B are changed into products C and D, the equilibrium constant is the ratio of product concentrations (i.e., multiplied by each other) over the concentrations of the reactants (i.e., multiplied by each other). The equilibrium constant is the ratio of product concentrations (i.e., multiplied by each other) over the concentrations of the reactants (i.e., multiplied by each other).

$$K_{eq} = ([C][D])/([A][B])$$

Le Châtelier's Principle

Just a few years after Guldberg and Waage described the law of mass action, the French chemist Henri Le Châtelier discovered a way to predict how condition changes affected chemical equilibrium. If a chemical system at equilibrium changes concentration, temperature, volume, or partial pressure, then the equilibrium must re-balance the changes to gain a new equilibrium.

In industry, Le Châtelier's principle is used to affect reversible chemical reactions. If an amount of product is removed from a reaction, then reactants will continue being converted to product. Higher product yields equal higher profit margins and less waste. In pharmacology, a high drug concentration will force increased binding to a particular receptor and alters biological activity.

> **Le Châtelier's principle** states that when a system at equilibrium is altered, the balanced equilibrium will decrease or counteract the disturbing effect.

Additional examples of what Le Châtelier's principle predicts include

- When a reactant's concentration increases, the equilibrium changes to make use of the added reactants by producing more products.

- When pressure on an equilibrium system increases, the equilibrium swings to reduce the pressure.

- When the volume of an equilibrium system (e.g., gas) is condensed and pressure increased, then the equilibrium will change to increase the volume and lower pressure.

- When the temperature of an *endothermic reaction* (e.g., heat absorbed from its surroundings) rises, the equilibrium will take up the extra heat by creating more products.

- When the temperature of an *exothermic reaction* (e.g., heat released into the environment) increases, equilibrium will shift to use up heat by creating more reactants.

Why Is Thermodynamics So Important?

Newton's laws of physics only apply under specific conditions, but the laws of thermodynamics have no exceptions. They describe all interactions in the universe. Energy is conserved, regardless of type or amount. Closed systems, whether at the nano or cosmic scale, always go towards greater entropy, unless impacted by an external force.

QUIZ

1. When a system at equilibrium is impacted and initial equilibrium changed, to lessen or counteract the effect of the disturbance, it is known as

   A. Hess's law
   B. Einstein's theory of relativity
   C. Le Châtelier's principle
   D. Williams's rule

2. In the equation, $R = k \times [A]^a \times [B]^b \times [C]^c$, what do A, B, and C stand for?

   A. Rate constant
   B. Concentrated reactants
   C. Temperature change
   D. Diluted reactants

3. The first law of thermodynamics explains that

   A. to every action, there is an equal and opposite reaction.
   B. at constant pressure, the volume of a confined gas is directly proportional to the absolute temperature.
   C. there is a season and a time for every purpose under heaven.
   D. matter/energy is neither created nor destroyed.

4. In thermodynamics, internal energy is made up of kinetic energy and

   A. potential energy.
   B. pressure.
   C. electricity.
   D. energy of motion.

5. The measure of unavailable energy within a closed system (e.g., the universe) is

   A. equilibrium.
   B. enthalpy.
   C. entropy.
   D. entomology.

6. Which equation would be used to measure a change in enthalpy at constant pressure?

   A. $\Delta H = \Delta H_{(products)} - \Delta H_{(reactants)}$
   B. $\Delta H = \Delta H_{(reactants)} - \Delta H_{(products)}$
   C. $R = k \times [A]^a \times [B]^b \times [C]^c$
   D. $K_{equililbium} = k_{forward\ reaction} / k_{reverse\ reaction}$

7. **Molar heat capacity describes the heat needed to**

    A. increase the temperature of 1 gram of a sample by 1°C.

    B. decrease the temperature of 1 mole of sample by 10°C.

    C. raise the temperature of 1 mole of a sample by 1°C.

    D. raise the temperature of a sample by 1°C.

8. **The amount of heat given off or absorbed in a chemical reaction is written as**

    A. ΔS.

    B. T.

    C. ΔG.

    D. ΔH.

9. **Which of the following does not usually influence the rate at which a chemical reaction takes place?**

    A. Temperature

    B. Skill of the researcher

    C. Catalyst

    D. Rate laws

10. **The law of mass action explains how a**

    A. condensation of a vapor occurs in a closed container.

    B. only applies when temperature and pressure are rising rapidly.

    C. a chemical's reaction rate is proportional to the sum of its reactant concentrations.

    D. is the basic principle behind mud slides after heavy rains.

chapter 12

Organic Chemistry: All about Carbon

Organic chemistry is all about the study of carbon. In this chapter, you will learn the naming, structure, bonding, and reactions of this amazing element.

CHAPTER OBJECTIVES

In this chapter, you will

- Learn about carbon and its connection to hydrogen
- Find out how carbon forms single, double, and triple bonds
- Discover the importance of functional groups
- Understand the difference between structural isomers
- Learn about benzene and aromatic hydrocarbons

What Is Organic Chemistry?

The study of compounds containing carbon (e.g., living things like orchids, algae, red blood cells, gnats, and elephants) is known as organic chemistry. Over 95% of all known chemicals contain carbon. At last count, there were over two million known organic compounds, nearly 20 times more than all the known chemicals combined.

> **Organic chemistry** is the chemistry of carbon and carbon-based compounds.

Organic compounds are also the largest group of covalently bonded compounds. They include everything from biologicals to petroleum-based chemicals, plastics, and synthetic fibers. Petroleum, also known as crude oil, is made up of decomposed organic compounds from plant and animal remains millions of years old.

At one time, it was thought organic compounds contained some type of "vital force," because they were once living organisms. However, in 1828, Friedrich Wohler used an *inorganic* salt, ammonium carbonate, to make urea (CH_4ON_2) in his laboratory. Urea is an organic molecule found in the urine of mammals. It was the first time an organic compound had been created from inorganic materials. This started chemists thinking about the possibilities of organic molecules in a new way.

Carbon—More Amazing Than Ever

Carbon's six electrons make up its 1s, 2s, and 2p orbitals (see Chap. 7). Its four valence electrons need four more electrons to complete an octet and stabilize the carbon atom. Extra electrons come from bonding with other atoms, including other carbons.

Although other elements form covalent bonds, carbon forms super stable carbon-carbon and carbon-hydrogen bonds. The huge variety of carbon-containing compounds in nature comes from carbon's ability to create stable bonds with other elements. Carbon forms long chain molecules like decane ($C_{10}H_{22}$), branching macromolecules like natural rubber, and ring structures like menthol (from peppermint). Figure 12-1 illustrates common organic molecules.

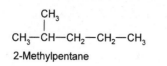

2-Methylpentane

$CH_3CH_2-O-CH_2CH_3$

Diethyl ether

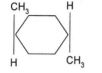

trans-1,4-Dimethylcyclohexane

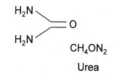

CH_4ON_2

Urea

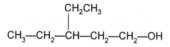

3-Ethyl-1-pentanol

FIGURE 12-1 · These organic molecules give you an idea of the diversity possible in organic chemistry.

? Still Struggling

In chemistry, a group (i.e., a family) is shown by a vertical column in the Periodic Table of elements. Family members are alike in electron configurations, especially the outermost shells, and have related chemical behavior. There are 18 groups in the standard periodic table. [Remember, a period is a horizontal listing of elements across the Periodic Table.] There are seven periods in all. Refer to the Periodic Table in Chap. 3 to check on periods and groups.

Silicon is a lot like carbon in atomic structure and forms silicon-silicon covalent bonds. But even though silicon and carbon are in the same family (group 14 of the Periodic Table), they don't act the same way. Silicon is from the third period and much bigger. In fact, silicon is over double carbon's size, but has longer and weaker bonds.

Picture a bridge between two river banks. A bridge across a 4-meter (12-foot)-wide stream is much stronger and more stable than one across an 8-meter (28-foot) stream, when the middle is not supported. Silicon simply can't do what stronger carbon molecules can.

Hydrocarbons

Organic compounds called *hydrocarbons*, are made up of, you guessed it, molecules with only carbon and hydrogen. Hydrocarbons are probably the easiest molecules in all of chemistry to learn. Once you get the basics, the rest is a matter of plugging in additional groups.

What Are Alkanes, Alkenes, and Alkynes?

Hydrocarbons are divided into subgroups depending on how carbon and hydrogen bond. There are three types: *alkanes*, *alkenes*, and *alkynes*. These classes are a *homologous series* of organic compounds, because they are either very much alike or react nearly the same.

Alkanes form the most bonds to different atoms, so they are called *saturated* molecules. They are stable when fully bonded. When double bonds are formed, hydrogen atoms are lost. Saturated fats, for example, have the maximum number of hydrogens that can be bonded to carbon (i.e., saturated with hydrogen atoms).

This is not the case for alkenes and alkynes. These are *unsaturated* molecules. Similarly, when talking about fats, saturation is all about the bonds in the fat molecules. (More in Chap. 13.) For now, let's take a look at how these carbon molecules are the same and what makes them different.

Alkanes

Hydrocarbons are called *alkanes* when they form only *single bonds* between carbon atoms. Methane (CH_4) and ethane (C_2H_6) are two of the simplest alkanes. While methane is for heating and cooking, ethane is mostly used to make ethene, an important industrial chemical.

Methane has only one carbon atom, so its carbon bonds to four separate hydrogens. These bonds face different directions to form a four-sided tetrahedron. The angles between the bonds are about 109°. Figure 12-2 shows the bond angles of the methane molecule.

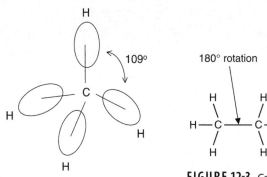

FIGURE 12-2 · The bond angles of 109° each are shown in the methane molecule.

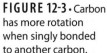

FIGURE 12-3 · Carbon has more rotation when singly bonded to another carbon.

Ethane has two carbons joined by a single bond. Each of carbon's three remaining bonds are bound to a hydrogen atom making a total of four bonds. Figure 12-3 shows the rotation about the carbon-carbon bond.

Alkenes

Hydrocarbons, known as *alkenes*, have a double bond between two carbons in the molecule. Figure 12-4 gives some examples of double-bonded carbons. Ethene or ethylene (C_2H_4) is an important industrial chemical used to make many different molecules, including plastics, such as poly*ethylene*.

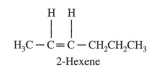

2-Hexene

CH$_3$—C=CHCH$_3$ with CH$_3$ branch
2-Methyl-2-butene

Ethene

FIGURE 12-4 · The double bond can have different elements and groups attached to the carbon.

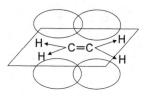

FIGURE 12-5 · The electrons are shared in the ethene molecule in a flat plane configuration.

Ethene is a simple double-bonded carbon molecule. It has two carbons with two bonds between them. Instead of sharing a single pair of electrons, they have two pairs of electrons. The remaining carbon bonds are to hydrogen atoms. In other words, each carbon has two bonds to the same carbon and two bonds to different hydrogens to make four bonds in all. Figure 12-5 shows how ethene shares its electrons.

Double bonds also hold molecules into rigid shapes. Unlike single bonds, no rotation or twisting is possible around the double bond. This inflexible carbon arrangement limits possible molecular shapes and chemical reactions.

Alkynes

Hydrocarbons, called *alkynes*, have a *triple bond* between two carbons in a molecule. Acetylene (C_2H_2) is the simplest alkyne and used to name related chemicals containing triple bonds, such as dimethylacetylene. Figure 12-6 shows some alkyne molecule structures.

$$H_3C-C\equiv C-H$$

Propyne (methylacetylene)

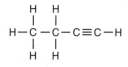

1-Butyne (ethylacetylene)

FIGURE 12-6 · Alkynes are triple-bonded carbon compounds.

Acetylene is also called ethyne. Ethyne has many industry uses (e.g., acetylene fuel for welding torches). Its two carbons form a triple bond and share three pairs of electrons. The remaining bonds are to hydrogen. In other words, each carbon has three bonds to the same carbon and one bond to a hydrogen to make four bonds total. Like double bonds, a triple bond allows no rotation between the carbons, although they can turn together like wheels on a car.

Ring Structures

Although carbons bond in straight and branched hydrocarbons, they can also form bonds in a variety of ring structures. The most stable structure has six carbons in a hexagonal shape. When bonds between ring carbons are all single, a molecule is called cyclohexane (C_6H_{12}). When the bonds alternate between single and double bonds, the molecule is benzene (C_6H_6). Figure 12-7 shows different functional groups positioned around benzene rings.

Benzene is part of a whole class of compounds called *aromatic* hydrocarbons. Many of these compounds have recognizable smells. Oil of vanilla or *vanillin* ($C_8H_8O_3$), extracted from the fermented seed pods of the vanilla orchid is

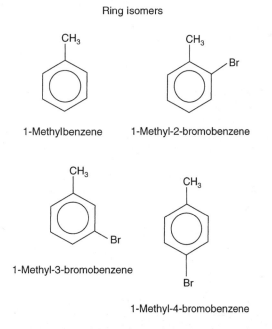

FIGURE 12-7 · Aromatic compounds vary according to the functional groups bonded around the rings.

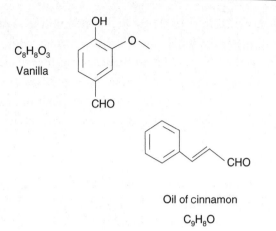

$C_8H_8O_3$
Vanilla

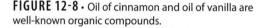

Oil of cinnamon
C_9H_8O

FIGURE 12-8 · Oil of cinnamon and oil of vanilla are well-known organic compounds.

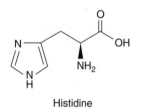

Histidine

FIGURE 12-9 · Allergy sufferers are fairly familiar with histidine.

shown in Figure 12-8. Another common aromatic compound, *cinnamaldehyde* or oil of cinnamon (C_9H_8O), obtained from the steam distillation of cinnamon tree bark is also shown.

Imidazoles ($C_3H_4N_2$) are another class of organic compounds with ring structures. Their structures have functional groups bonded in different places around a ring. These ring structures are part of biological molecules (e.g., histidine and histamine hormones) involved in allergic reactions caused by high pollen count. Figure 12-9 shows the histidine structure.

Naming Organics

Just as scientists needed a common language for the elements (refer back to Chap. 3), organic chemists use standardized naming guidelines. The International Union of Pure and Applied Chemistry (IUPAC) method of naming hydrocarbons is fairly simple. Each name has two parts, a prefix and a suffix, which describes the type of molecule.

The prefix is based on the number of carbons in the molecule. Some of the basic prefixes (i.e., with carbon number in parentheses) include: meth- (1), eth- (2), prop- (3), but- (4), pent- (5), hex- (6), hep- (7), oct- (8), non- (9), and dec- (10).

The organic suffix names the molecule class. All alkanes end in *-ane*. All alkenes end in *-ene*. All alkynes end in *-yne*. If you know these, you'll breeze through naming in organic chemistry.

EXAMPLE 12-1
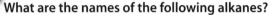

What are the names of the following alkanes?

(1) CH_4
(2) C_3H_8
(3) C_6H_{14}
(4) C_9H_{20}

SOLUTION

Did you get (1) methane, (2) propane, (3) hexane, and (4) nonane?

Organic molecules aren't limited to simple structures. They have many different groups attached to them. These groups are included in the name as well. When discussing hydrocarbon groups, the same prefixes are used with "*yl*" added at the end. For example, methane, then, becomes *methyl*, ethane becomes *ethyl*, and so on. These are *alkyl* groups. The same rules apply to *alkenyl* and *alkynyl* groups.

When alkanes, alkenes, and alkynes have different bond types, molecule naming has a few more IUPAC rules.

1. The molecule's base name comes from the longest chain with a double or triple carbon bond. For example, if the longest chain has four carbons and a triple bond, then the base name would be butyne.

2. When naming different molecule parts, numbers are used to locate the different groups (e.g., methyl groups) and double bonds. Numbering is based upon bond order (e.g., the first position belongs to the carbon with bonds to atoms with the highest atomic number) and numbering continues until the end of the base molecule is reached.

3. These rules also apply when numbering six carbon ring molecules.

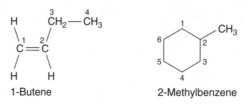

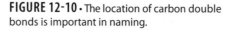

1-Butene   2-Methylbenzene

FIGURE 12-10 · The location of carbon double bonds is important in naming.

? Still Struggling

Look at 1-butene (CH_2=$CHCH_2CH_3$). Why is its name not 3-butene? At one end of the molecule, the carbon has two bonds to carbon and two to hydrogen. At the other end, the carbon has only one bond to carbon and three to hydrogen. The carbon with more carbon bonds gets top billing and is number 1. In another example, 2-methylbenzene has a methyl (CH_3) attached at the second carbon in the ring. Figure 12-10 shows these structures.

Bond Polarity

When two atoms share electrons in a covalent bond, the relationship may or may not be equal. When electrons are shared by the same elements, such as two carbon atoms, they are shared equally. The electron distribution is symmetrical.

This is not always the case, however. If the atoms of two or more different elements share a pair of electrons, the electron density at one or another of the atoms will be different. Some atoms have a stronger attraction for a shared electron pair than another. (Look back to Chap. 8.)

> **Bond polarity** takes place when electron pairs are unequally shared between atoms of different elements.

The bonding is not equal in the general sharing (one atom gets more time with the electron pair than the other). This unequal sharing causes a charge shift to one atom more than the other.

Still Struggling

It's important to remember that some molecules have polar bonds, but a molecule as a *whole* is not polar. This is easier to understand when polar bonds face in opposite directions. They cancel each other out, such as two equally strong people pulling in opposite directions on the same rope.

In carbon tetrachloride, (CCl_4), there are four very polar bonds, but the bond angles and symmetry causes the bond dipoles to cancel out. Carbon tetrachloride, then, is a nonpolar molecule. (Take a look again at Fig. 12-2.)

Unequal sharing or stronger/weaker attraction of shared electrons explains why some molecules have certain reactivity with some atoms and very different reactivity with others. Melting and boiling points, as well as the ability to form higher energy compounds, depend on these polarization effects.

Common Functional Groups

Organic chemistry is not just about carbon and hydrogen bonds. Carbon bonds with many other elements to create all kinds of interesting molecules. These bonds introduce new properties (i.e., functions) to molecules, and so they are called *functional groups*.

Bond polarity and the properties of non-carbon elements play an important role in the properties of functional groups. Figure 12-11 shows functional

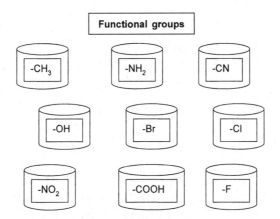

FIGURE 12-11 · A wide variety of functional groups combine with carbon to produce organic molecules.

groups that combine with carbon to make different organic molecules. These molecules include alcohols, phenols, and carboxylic acids (–OH), as well as aldehydes, ketones, esters, and anhydrides (R–C=O) (C=O) when carbon combines with oxygen and other functional groups. The bonding between carbon and nitrogen results in amines (N–H) and nitriles (C≡N).

Each functional group makes a molecule behave in a similar way so the functional groups are another way to classify and talk about organic chemicals in a homologous series.

To highlight added functional R is commonly used to stand for a carbon group (e.g., CH_3-, CH_3CH_2-, $CH_3CH_2CH_2-$, or CH_3CHCH_3-). This notation makes finding a functional group a lot easier and highlights its role in a molecule's properties. We'll learn more about key functional groups of biological systems in the next chapter on biochemistry.

EXAMPLE 12-2

The following shorthand formulas show how a few functional groups are added to the base carbon group (R):

- Ethanol is CH_3CH_2OH or R–OH
- Acetic acid (carboxylic acid commonly known as vinegar) is CH_3CO_2H or R–COOH
- Formaldehyde (methanal) is R–HC=O
- Methylamine is CH_3NH_2 or R–NH$_2$
- Benzylchloride is a benzene ring–CH_2Cl

Isomers

Jöns Jacob Berzelius was a Swedish chemist who first used symbols to name elements and use the word *isomer* in his work on chemical atomic weights. Isomers have the same molecular formula, but different structural formulas. There are two kinds of isomers—*structural isomers* and *stereoisomers* (spatial isomers). [Note: Be careful not to confuse isomers with isotopes (i.e., element properties, not molecules.)] (More about nuclear chemistry and isotopes in Chap. 14.)

Isomers have the same molecular formulas, but different three-dimensional structures.

Still Struggling

Think of yourself as a baker and molecules as your dough. The same dough recipe can be used to make loaves, ropes, twists, rolls, or a dozen other shapes. In the same way, isomers can take on different forms. The dough (formula) is the same, but the different isomeric shapes may cause them to react very differently. If the dough is made into a long, thin rope, and cooked on a heat-conducting metal baking sheet, it will bake much faster than a round, solid core of dough baked in a thick-walled clay baker. When both shapes are cooked at the same temperature for the same time period, the thin rope will cook (and probably burn) in half the time it takes for the round loaf to bake. Shape makes a lot of difference in how similar formulas react.

Structural Isomers

Structural isomers have the same molecular formulas, but the atoms are connected differently in each molecule. When the atoms are connected into one chain, a molecule can be quite long. When the atoms are connected into branches, the molecule can be a lot shorter. Structural isomers of pentane (C_5H_{12}) are shown in Fig. 12-12. Their groups are arranged differently in structure, but their formulas are the same.

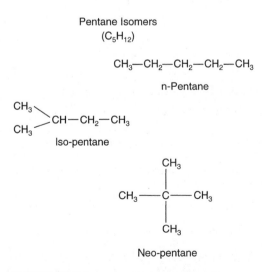

Pentane Isomers
(C_5H_{12})

$$CH_3—CH_2—CH_2—CH_2—CH_3$$

n-Pentane

Iso-pentane

Neo-pentane

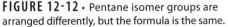

FIGURE 12-12 · Pentane isomer groups are arranged differently, but the formula is the same.

Stereoisomers

Stereoisomers have the same molecular formulas and arrangements of atoms, but the molecule's groups are positioned differently in their three-dimensional orientations in space. Stereoisomers are not structural isomers. Their atoms and functional groups are linked in the same order, but are located in different arrangements in space depending on their size and types of bonding.

Cis–Trans Isomerism

A subclass of stereoisomers includes *cis–trans isomers* or *geometric isomers*. This stereoisomer type is twisted through a limited rotation around a double bond or ring system.

These molecules focus around the location of groups across a double bond. If the groups are on the same side of the double bond, it is known as a *cis* formation. However, if the groups are located on opposite sides of the double bond (diagonally) from each other, they are in a *trans* formation. In Fig. 12-13 the *cis–trans* structures of dichloroethane are shown. In one view (*cis*), both chlorines are on one side of the double bond, while in the other view (*trans*), they are on opposite sides.

Enantiomers

Another subclass of stereoisomers is known as *enantiomers*. These isomers are mirror images of each other and are not superimposable. Figure 12-14 shows the non-superimposable isomers of chlorobromoiodomethane. This property is called *chirality* (for chiral molecules). The word *chiral* comes from the Greek word *cheir*, meaning "hand."

When the mirror image of a molecule is not identical to itself, this is thought of as "handedness" or a chiral arrangement. Chiral molecules interact with light

Cis/trans isomers

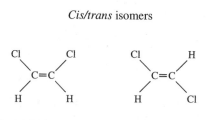

cis –1,2-Dichloroethene   *trans* –1,2-Dichloroethene

FIGURE 12-13 · *Cis/trans* bonding of dichloro-ethane shows the placement of chlorine atoms.

2-Butanol

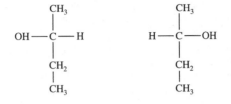

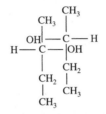

FIGURE 12-14 · Chirality is a factor in studying stereoisomers.

differently, and are also called *optical isomers*. (Note: molecules or objects that *are* superimposable are called *achiral*.)

EXAMPLE 12-3

Which of the following object(s) are chiral (c)? Which are achiral (a)?

(1) Fork
(2) Ear
(3) Hat
(4) Spiral staircase
(5) Shoe
(6) Baseball bat

SOLUTION

Did you get: (1) a, (2) c, (3) a, (4) c, (5) c, and (6) a?

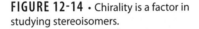

Still Struggling

To illustrate chirality, take a pencil and outline your right hand. Now do the same thing with your left hand on another piece of paper. If you lay the right outline on top of the left outline, they aren't identical. Try it!

Organic Reactions

Organic reactions can be grouped into general types where organic substances are created or joined.

Addition Reactions

When compounds contain double and triple bonds, the most common element interactions take place as *addition reactions*. For example, when a double-bonded hydrocarbon like ethene ($CH_2=CH_2$) is added to another compound with single bonds, the elements of the new compound position themselves on either side of the double bond. The double bond is broken and single bonds are formed.

EXAMPLE 12-4

The addition of hydrogen chloride to ethene yields chloroethane:

$$CH_2=CH_2 + HCl \rightarrow H_3-C-C-H_2Cl \rightarrow CH_3CH_2Cl$$

When diatomic chloride is added to ethene, then the result is 1,2-dichloroethane:

$$CH_2=CH_2 + Cl_2 \rightarrow ClH_2-C-C-H_2Cl \rightarrow CH_2CH_2Cl_2$$

For triple-bonded hydrocarbons, the addition of bromine to ethyne would yield:

$$H-C\equiv C-H + 2\,Br_2 \rightarrow Br_2H-C-C-HBr_2 \rightarrow CH_2Br_4$$

Substitution Reactions

Hydrocarbons react with other elements in specific ways. The hydrogen bond breaks and the reacting element slips into place. For example, methane's reaction with members of the halogen group (e.g., fluorine, chlorine, or bromine), creates halomethanes and hydrogen halides.

EXAMPLE 12-5

This example shows possible substitution reactions between a hydrocarbon (methane, CH_4) and a halogen (chlorine, Cl):

$$CH_4 + Cl_2 \rightarrow CH_3Cl + HCl$$

The same reaction takes place with a halogen added to the singly substituted CH_3Cl:

$$CH_3Cl + Cl_2 \rightarrow CH_2Cl_2 + HCl$$

$$CH_2Cl_2 + Cl_2 \rightarrow CHCl_3 + HCl$$

$$CHCl_3 + Cl_2 \rightarrow CCl_4 + HCl$$

Substitution reactions are simple and very common among organic compounds, but can take place between other elements too.

EXAMPLE 12-6

A substitution reaction between fluoromethane (CH_3F) and potassium bromide (KBr) trades elements in the following reaction:

$$CH_3F + KBr \rightarrow CH_3Br + KF$$

Polymerization and Cracking

When two smaller compounds combine to form a much bigger compound, it is known as *polymerization*. The opposite process takes place when a larger compound is broken down into smaller compounds. This is known as *cracking*. Cracking is used by chemical engineers when crude oil (made up of hydrocarbons) is separated and refined using high pressure and temperature or a lower temperature and pressure in the presence of a catalyst.

EXAMPLE 12-7

This example shows compounds produced after cracking a large hydrocarbon:

$$C_{15}H_{32} \rightarrow 2\ (C_2H_4)\ (\text{ethane}) + C_3H_6\ (\text{propene}) + C_8H_{18}\ (\text{octane})$$

Oxidation

Organic compounds can also react with an oxidizing element (e.g., oxygen or fluorine) at high temperatures to make carbon dioxide (CO_2) and water (H_2O) (i.e., combustion). The combustion of hydrogen and oxygen, a reaction used in rocket engines is shown.

$$2H_2 + O_2 \rightarrow 2H_2O_{(vapor)} + \text{heat energy}$$

Esterification

For *esterification* to happen, (i.e., making esters R–COOR′; R and R′ are alkyl or aryl groups) you need heat. A carboxylic acid (e.g., R–CO–OH) is mixed with an alcohol (R′–OH), and an acid catalyst is used to speed up the reaction. The chemical structure of the alcohol (e.g., methanol (CH_3OH), acid [e.g., acetic acid or vinegar (CH_3CO_2H)] and acid catalyst [e.g., hydrochloric acid, (HCl)] all effect the esterification reaction rate. More about esters in the next chapter.

Fermentation

The chemical process of *fermentation* has been in existence for thousands of years. Ancient Egyptians brewed beer from organic foodstuffs (e.g., wheat, malt, and yeast). Then, they fermented the mixture without oxygen for several days to weeks.

> **Fermentation** is the anaerobic (no oxygen) conversion of sugar from grain into carbon dioxide and alcohol through reaction with yeast.

Fermentation creates lactate, acetic acid, ethanol, and other simple products from more complex organic compounds.

The organic structures and bonding described in this chapter should give you a broad overview of the kinds of reactions that organic molecules are involved in.

QUIZ

1. Carbon-based compounds are the main components of
   A. inorganic chemistry.
   B. polymerization.
   C. organic chemistry.
   D. particle-wave chemistry.

2. Acetic acid (commonly known as vinegar) has which of the following functional groups added to its carbon chain (R–)?
   A. R–OH
   B. R–HC=O
   C. R–NH_2
   D. R–COOH

3. Pentane is a hydrocarbon molecule with how many carbons?
   A. 3
   B. 5
   C. 7
   D. 9

4. Ethane (CH_3CH_3) reacts with bromine vapor in light to form bromoethane (CH_3CH_2Br) and
   A. 1-bromobutane [$CH_3(CH_2)_2CH_2Br$].
   B. sec-butyl bromide (C_4H_9Br).
   C. hydrogen bromide (HBr).
   D. n-propyl bromide (C_3H_7Br).

5. The nonpolar compound, carbon tetraflouride, has bond angles of
   A. 25°.
   B. 60°.
   C. 90°.
   D. 109°.

6. When electron pairs are unequally shared between atoms of different elements, it is known as
   A. bond polarity.
   B. a homologous series.
   C. an enantiomer.
   D. an isocline.

7. **2-Methylbenzene has a methyl (CH$_3$) group attached at which carbon in the ring?**

   A. First

   B. Second

   C. Third

   D. Fourth

8. **Cinnamaldehyde, also written as 3-phenyl-2-propenal, has which of the following formulas?**

   A. C_2H_5OH

   B. $C_6H_5CH_2NH_2$

   C. C_9H_8O

   D. $C_3H_5(OH)_3$

9. **Two mirror image molecules that are superimposable are**

   A. chiral.

   B. achiral.

   C. radioactive.

   D. only available to form triple bonds.

10. **When ethanol (CH$_3$CH$_2$OH) reacts with hydrogen iodide (HI) to form iodoethane (ethyl iodide) and water, the OH is replaced by**

   A. two hydrogen atoms.

   B. a carboxylic acid.

   C. an amine group.

   D. an iodine atom.

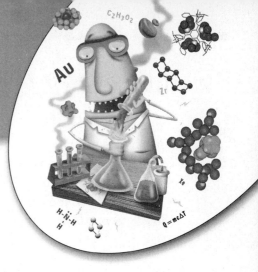

chapter 13

Biochemistry

Chemistry is life! Wouldn't it be great if this was on shirts and billboards across the country? After all, elements make up everything we know to exist in the world. People seeking jobs involving chemistry or related sciences are especially happy since there are as many applications of chemical knowledge as there are disciplines of study. In this chapter, the diverse molecules of living organisms and systems will be described along with important properties and major reactions.

CHAPTER OBJECTIVES

In this chapter, you will

- Understand the difference between hydrophilic and hydrophobic molecules
- Become skilled at naming different kinds of biological molecules
- Learn the importance of carboxylic acids
- Find out about the different classes of amines
- Discover the impact of enzymes on reactions
- Learn the difference between saturated and unsaturated fats

What Is Biochemistry?

Biochemistry, describes the chemistry of living systems. This overlaps with organic chemistry since most compounds in living cells contain carbon. For example, a single-celled organism, through its *metabolism*, builds up or breaks down organic molecules all through its life cycle.

The biological molecules in living cells, organs, systems, and the environment are divided into four types: *proteins, carbohydrates, nucleic acids,* and *lipids.* Figure 13-1 shows the variety of biochemical molecules found in nature. Most of these molecules are simple structures covalently bonded to similar molecules, but some are huge compounds. These are called *macromolecules.* Figure 13-2 shows the structure of the macromolecule β-carotene (i.e., orange color in carrots).

trans-1, 4-Dimethylcyclohexan

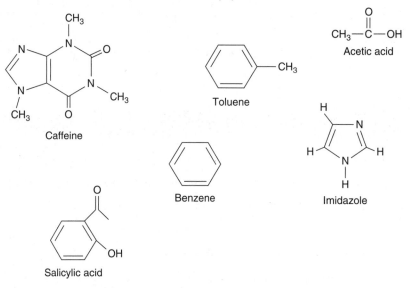

FIGURE 13-1 · A few of the biological molecules found in nature.

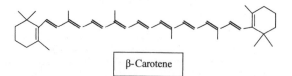

β-Carotene

FIGURE 13-2 · The structure of β-carotene shows the complex nature of proteins.

Hydrocarbons—Hydrophilic versus Hydrophobic

In the last chapter, we learned how different types of organic compounds form by binding functional groups to carbon. Table 13-1 lists several of these groups. In living systems, these groups have certain properties.

We know polar molecules are attracted to water molecules because of bonding configuration. These are *hydrophilic* or "water loving" since they interact with water by forming hydrogen bonds. Nonpolar molecules, on the other hand, are *hydrophobic* or "water fearing" and don't dissolve in water. However, there is more to it than that. Large molecules can have portions that are hydrophobic while other sections are hydrophilic. Figure 13-3 shows hydrophilic and hydrophobic molecules.

TABLE 13-1 Many different organic compounds are formed by adding functional groups to carbon

Name	Functional Group	Example Molecule	Formula
Alcohol	R–OH	Ethanol	CH_3CH_2OH
Aldehyde	R–HC=O	Propionaldehyde	CH_3CH_2CHO
Amide	$R–NH_2–C=O$	Acetamide	CH_3CONH_2
1°Amine	$R–NH_2$	Propylamine	$CH_3CH_2CH_2NH_2$
2°Amine	R–HN–R	Dimethyl amine	CH_3NHCH_3
Carboxylic Acid	R–OOH	Acetic acid	CH_3CO_2H
Ester	RCOOR′	Methyl propanoate	$CH_3CH_2COOCH_3$
Ether	R–O–R	Diethyl ether	$CH_3CH_2OCH_2CH_3$
Ketone	C–CO–C	Butanone (methylethyl ketone)	$CH_3COCH_2CH_3$
Nitrile	R–C≡N	Ethyl cyanide (cyanoethane)	CH_3CH_2CN
Phenol	R	Hydroxybenzene (carbolic acid)	C_6H_5OH

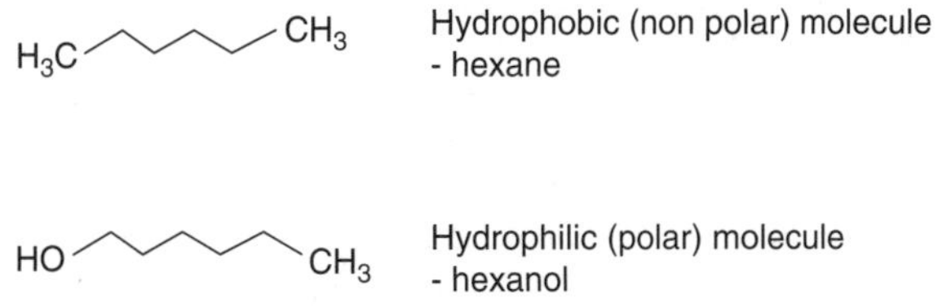

H_3C ⌇⌇⌇⌇ CH_3 Hydrophobic (non polar) molecule - hexane

HO ⌇⌇⌇⌇ CH_3 Hydrophilic (polar) molecule - hexanol

FIGURE 13-3 • Hydrophilic and hydrophobic molecules can be very similar.

Surfactants

A *surfactant* (i.e., short for SURFace ACTive AgeNT) is a substance with a hydrophilic head and hydrophobic tail that repels water and binds to oil. Surfactants help keep particles mixed instead of letting them settle back into separate layers (e.g., oil and vinegar). These molecules can go either way. They are semisoluble in both organic and aqueous solvents.

> **Surfactants** are molecules with hydrophobic and hydrophilic components that lower surface tension.

? Still Struggling?

Detergents are a common type of surfactant. When added to a load of laundry in a washing machine, surfactants latch onto dirt and pull it away from a fabric's fibers. The suspended surfactant and dirt can then be removed with the rinse water, instead of settling back onto the clothes' surfaces.

Carboxylic Acids

Carboxylic acids are organic compounds containing a –COOH group bonded to a hydrogen atom or an alkyl group. Some familiar carboxylic acids include lactic acid in milk, citric acid in lemons, and oleic acid in olive oil.

Naming Carboxylic Acids

When naming carboxylic acids, the parent chain must have the carboxyl carbon in the number 1 position. Then, the name of the attached alkane is found by replacing the *-e* on the end with *-oic acid*.

▢ EXAMPLE 13-1

HCOOH (methane becomes methan*oic* acid)

CH_3COOH (ethane becomes ethanoic acid, also known as acetic acid)

$CH_3CH_2CH_2COOH$ (butane becomes butanoic acid)

The carboxyl group is also weakly acidic and carboxylic acids neutralize OH^- as in:

$RCOOH + NaOH \rightarrow RCOONa + H_2O$

Esters

Esters, another group of organic compounds, are everywhere. They form by combining an acid and an alcohol, while losing a molecule of water in the process. Carboxylic acids are often used to create esters.

The general chemical formulas for esters are written as RCO_2R', where R and R' are the organic groups of the carboxylic acid and alcohol correspondingly.

These molecules, found in fruits and flowers, give off recognizable smells (e.g., wintergreen, strawberry). Because they often smell good, esters are used in essential oils and fragrances.

Different esters, *pheromones*, are secreted by organisms (e.g., ants and bees) to attract mates, signal danger, or pass along directions. Still other nitrate esters, such as nitroglycerin, are extremely explosive. Table 13-2 lists different esters.

▢ EXAMPLE 13-2

Butanol [$CH_3(CH_2)_3OH$] + acetic acid (CH_3COOH) → butyl acetate
($CH_3CO_2C_4H_9$)

TABLE 13-2 Esters give fruits and other foods good smells	
Ester	**Scent**
$CH_3CH_2CH_2COOCH_3$	Apple
$CH_3(CH_2)_2COO(CH_2)_4CH_3$	Apricot
$CH_3COO(CH_2)_4CH_3$	Banana
$CH_3COOCH_2CH_3$	Jasmine
$CH_3COO(CH_2)_7CH_3$	Orange
$CH_3COO(CH_2)_2CH(CH_3)_2$	Pear
$CH_3(CH_2)_2COOCH_2CH_3$	Pineapple
$HCOOCH_2CH(CH_3)_2$	Raspberry
$HCOOCH_2CH_3$	Rum
$CH_3CH_2CH_2COOCH_2CH_3$	Strawberry
$C_6H_4(HO)COOCH_3$	Wintergreen

Naming Esters

To name an ester, first find the alkyl group attached to the oxygen atom, then number from the end closest to the –CO– group. Find the alkane linking the carbon atoms together. If a continuous link of carbon atoms is interrupted by an oxygen atom, individually name the two alkanes before and after the oxygen atom. The longer structural alkane should contain the carbonyl atom. Lastly, change the parent chain -e ending and replace it with an -oate.

In short, naming should follow the pattern below:

Alkane farthest from carbonyl ➔ alkane closest to carbonyl ➔ parent chain

? Still Struggling?

Look at the reaction: ethanoic acid + ethanol ➔ ethyl ethanoate

$$CH_3COOH + CH_2CH_3OH \rightarrow CH_3COOCH_2CH_3 + H_2O$$

A hydrogen from the acid combines with another hydrogen and oxygen from the acid to form water and the ester.

Amines

Within *amines*, the simplest being ammonia (NH_3), hydrogen atoms are replaced by functional groups. If you remember from Table 13-1, these amines have the general formula RNH_2, where R represents an alkyl group.

> Generally, when naming **amines**, find the alkyl group and then add the suffix *-amine* to its name.

Classes of Amines (1°, 2°, 3°)

Primary Amines

Primary amines (1°) have one hydrocarbon group bonded to an *amino* group ($-NH_2$).

EXAMPLE 13-3

Look at the following amines. Can you name them?

(1) CH_3-NH_2

(2) $CH_3-CH_2-NH_2$

(3) $CH_3-CH_2-CH_2-NH_2$

✔ SOLUTION

Did you get: (1) methylamine, (2) ethylamine, and (3) propylamine?

Naming primary amines is fairly straightforward, but when adding a $-NH_2$ group to a chain it is often easier to name the compound by using the "amino" form.

EXAMPLE 13-4

Here are some compounds named from the NH_2 group's location.

$$CH_3-CH_2-CH_2-NH_2 = \text{1-aminopropane}$$

$$\begin{array}{c} NH_2 \\ | \\ CH_3-CH-CH_3 \end{array} = \text{2-aminopropane}$$

Secondary Amines

In secondary (2°) amines, two hydrogen atoms of the ammonia molecule are replaced by hydrocarbon groups. In the following example, both hydrocarbon groups are alkyl groups and the same.

EXAMPLE 13-5

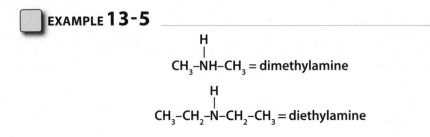

$$CH_3-NH-CH_3 = \text{dimethylamine}$$

$$CH_3-CH_2-N-CH_2-CH_3 = \text{diethylamine}$$

Tertiary Amines

In tertiary (3°) amines, all of ammonia's hydrogen atoms are replaced by hydrocarbon groups. Naming is the same as secondary amines.

EXAMPLE 13-6

$$CH_3- N-CH_3 = \text{trimethylamine}$$

EXAMPLE 13-7

The organic compound with the formula $C_9H_{21}N$ or $N(C(CH_3)_2H)_3$ is called triisopropylamine. As shown, the branched structure is crowded with methyl groups. It is a tertiary amine:

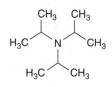

Amides

Amides are another compound formed from carboxylic acids, but instead of the –COOH group, in an amide the –OH group is replaced by an –NH_2 group (i.e., –$CONH_2$):

In other words, an amide is an organic compound with a nitrogen attached to a carbonyl R–C(=O)NR$_2$ as

Similar to secondary amines, a simple amide can be made from ammonia (i.e., when a hydrogen atom is replaced by an acyl group) and written as RC(O) NH$_2$. Amides, used in industry include, crayons, nylon, and inks.

Naming Amides

To name an amide, you add -*amide* to the parent acid's name from which it was formed. So, the simplest amide obtained from acetic acid is called *acetamide* (CH$_3$CONH$_2$). In a branched chain, the carbon of the –CONH$_2$ group counts as the first carbon.

Figure 13-4 shows the structures of acetamide (i.e., ethanamide), 3-methylbutanamide (branched), and urea.

In biological systems, amides form covalent linkages within long chain proteins. *Urea*, also known as carbamide, is the main nitrogen-containing compound, (NH$_2$)$_2$CO, in the urine of mammals. It is very important in nitrogen metabolism. We will learn more about the biochemicals of living systems in the next section.

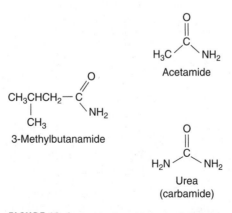

FIGURE 13-4 · Amides have nitrogen-containing functional groups.

Phenols

Phenols, or *phenolics*, are a class of organic compounds made up of one or more hydroxyl (–OH) groups bonded to an aromatic hydrocarbon. Phenol (also known as carbolic acid, benzophenol, or hydroxybenzene) is produced from benzene.

The chemical formula of phenol, C_6H_5OH, has a hydroxyl group bonded to a benzene ring.

Phenols are found in foods, human and animal wastes, decomposing organic material, and in coal tar and creosote. Salicylic acid is a natural phenolic compound found in willow bark and used by early people as a tea to cure headaches. Phenols are also formed in the atmosphere during forest fires by the breakdown of benzene. See Fig. 13-5 for more illustrations of phenol structures, including salicylic acid.

Naming Phenols

Phenol is a common name for hydroxybenzene. To name other phenols, just note the location of the functional group(s) attached to the benzene ring. Substituted

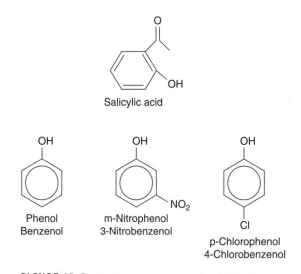

Salicylic acid

Phenol
Benzenol

m-Nitrophenol
3-Nitrobenzenol

p-Chlorophenol
4-Chlorobenzenol

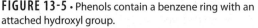

FIGURE 13-5 · Phenols contain a benzene ring with an attached hydroxyl group.

groups are either numbered or shown in an (1) ortho, (2) meta, or (3) para position from the –OH location. Whenever the parent molecule is called a phenol, the common system of nomenclature is being used.

A phenol's IUPAC (International Union of Pure and Applied Chemistry) name is *benzenol* to show that an –OH group has been attached to a ring structure. So, the parent molecule (benzenol), and any added groups, are always numbered with the –OH group being given the first position. (Refer again to the compounds in Fig. 13-5.) The top name is the common name and the second is the IUPAC name.

What Are Proteins?

Proteins are organic macromolecules containing carbon, hydrogen, oxygen, nitrogen, and often sulfur. These have one or more amino acid chains in a specific order like beads on a string. Proteins are essential to all living cells and include the enzymes, hormones, and antibodies needed to keep organisms healthy and functioning properly. They are important for the structure, function, growth, regulation, and repair of tissues. Proteins are also found in many foods (e.g., fish, meat, eggs, milk, and nuts).

> **Proteins** are made up of small molecules containing an amino group (–NH$_2$) and a carboxyl (–COOH) group.

Amino Acids

Amino acids, with the general structure R–CH(NH$_2$)COOH, make up proteins and are the building blocks of enzymes, hormones, proteins, and body tissues. Figure 13-6 shows the general structure of amino acids.

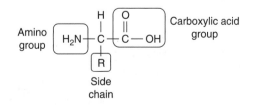

FIGURE 13-6 · Amino acids usually have three distinct functional groups.

A *peptide bond* [e.g., C(O)NH] is a covalent bond formed between two molecules where a carboxyl group from one molecule reacts with the amine group of another molecule. When this happens, a water molecule (H_2O) is released. This dehydration synthesis (i.e., reaction between amino acids) is also called a *condensation reaction*. A peptide bond is formed, and the resulting molecule is an amide. The four-atom functional group –C(=O)NH– is called a *peptide link*.

Peptides form compounds of different sizes.

- A simple *peptide* is a compound made up of two or more amino acids.
- An *oligopeptide* has 10 or less amino acids.
- *Polypeptides* and *proteins* are chains of 10 or more amino acids.

Protein reactions are made up of many different amino acid combinations reacting with water, salts, and other elements to create or enhance body functions. Amino acids contain a variety of nonprotein ions like metals (e.g., Zn^{2+}, Fe^{2+}, Mg^{2+}). For example, within the hemoglobin molecule, iron plays a critical part in transferring oxygen in living systems.

Other organic proteins serve different biological functions in organisms. Some offer structural strength (e.g., chitin shells of crabs), some provide transport (e.g., hemoglobin), some act as blueprints for cell and organ development (e.g., DNA), some serve as messengers (e.g., hormones) between body organs, and some speed up metabolic reactions (e.g., enzymes).

In fact, proteins are such multipurpose molecules, their molecular weights range from 6×10^3 to millions of atomic mass units.

Nucleic Acids

Nucleic acids include a group of large, complex, organic acids found in all living cells and viruses. They are made up of nucleotide chains that transfer genetic information. Nucleotides have three parts: a nitrogenous heterocyclic base (i.e., a purine or a pyrimidine), a pentose sugar, and a phosphate group. Various nucleic acids have different sugars in the chain.

Nucleic acids are single-stranded and double-stranded. RNA is usually single-stranded, but a strand often folds back upon itself to form double-helical areas.

A double-stranded nucleic acid is composed of two single-stranded nucleic acids hydrogen bonded together. DNA is usually double-stranded, though some viruses have single-stranded DNA as their genome (i.e., genetic blueprint).

Deoxyribonucleic Acid (DNA)

Nucleic acids are crucial components of protein synthesis, cell function regulation, and heredity. The key players, *deoxyribonucleic acid* (DNA) and *ribonucleic acid* (RNA), have long chains of repeating nucleotide bases, but with different structures and functions. DNA and RNA are made up of a series of *purine* and *pyrimidine* bases [e.g., adenine (A), guanine (G), cytosine (C), thymine (T), or uracil (RNA)].

By forming nucleotides with triplets of bases or *codons*, they form the genetic code of living organisms. When broken down, nucleic acids yield phosphoric acid, sugars, and an assortment of coded organic bases. Figure 13-7 shows how nitrogenous base pairs are arranged in the double helix configuration of DNA.

There are many fairly new biochemistry techniques that have become important in the study of heredity and genetics. In 1993, Kary Mullis received the Nobel Prize in Chemistry for the invention of a polymerase chain reaction technique for amplifying *deoxyribonucleic acid* (DNA). That same year, Canadian chemist Michael Smith also received a Nobel Prize in Chemistry for his method of splicing foreign gene segments, designed to modify the production of a specific protein, into another organism's DNA.

These advances opened the gates to research on designer proteins and molecules created for a specific purpose. Designer proteins are now used for improved insulin to treat diabetes, artificial fabrics to treat and protect burn patients, and industrial foams that clump and remove oil spills.

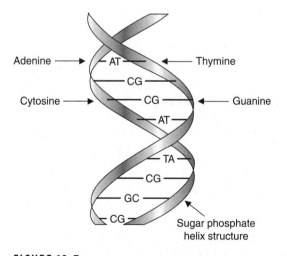

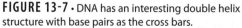

FIGURE 13-7 · DNA has an interesting double helix structure with base pairs as the cross bars.

What Are Enzymes?

Enzymes, the "movers and shakers" of a cell, are extremely important. They help build proteins, carry materials around a cell, and make biochemical reactions possible. Without enzymes, our body chemistry would stop working and life wouldn't exist.

Enzymes are proteins that lower the activation energy or *catalyze* roughly 4000 different biochemical reactions. Thousands of enzymes, produced by cells within the body, are created to speed up the reaction rate of biological processes. Like all catalysts, enzymes aren't used up during the reactions they catalyze, so very small amounts have a huge impact.

Biological enzymes are also much more specific than general catalysts. Each enzyme in the fluids and tissues of the body works on certain specific substrates. It's a lock and key situation; only the right key, opens the door. With the right enzyme bound to the right molecule (i.e., substrate), biological reactions take place converting molecules into products.

> **Enzyme specificity** describes how an enzyme's active site fits a certain substrate and reacts, while others don't fit and no reaction occurs.

The arrangement of an enzyme's *active site* (i.e., within an enzyme's structure where a substrate, functional group, or cofactor bind) is crucial to a reaction. Since a catalyst's active site and substrate must be an exact fit to start a reaction, some substrate molecules will react and others won't.

? Still Struggling?

When you get a paper cut and dab on hydrogen peroxide, it starts bubbling and turns white. The reason this happens is because an enzyme in the blood called *catalase* makes it happen. Catalase causes hydrogen peroxide (H_2O_2) to break down into hydrogen and oxygen. The chemical reaction is below.

$$2\,H_2O_2 = 2\,H_2O + O_2$$

The reaction is so exact that if you pour hydrogen peroxide on unbroken skin (i.e., without blood containing catalase) no bubbling takes place.

Cofactors and Coenzymes

Enzymes often have helper molecules called *cofactors* to accomplish specific tasks. These chemical compounds are tightly bound to an enzyme's active structure and are needed for enzymes to function.

A *coenzyme* is a small organic molecule that transports chemical groups between enzymes. Acetyl coenzyme A (i.e., acetyl-CoA) is an important metabolic molecule involved in lots of biochemical reactions. It helps move carbon atoms of an acetyl group within the Krebs cycle to become oxidized and make energy.

Enzymes make up various types according to the reactions they catalyze. These are

- *Oxidoreductases* catalyze the reduction or oxidation of a molecule.
- *Transferases* catalyze the movement of a group of atoms from one molecule to another.
- *Hydrolases* catalyze hydrolysis reactions and their reverse reactions.
- *Isomerases* catalyze the conversion of a molecule into an isomer.
- *Lyases* act to add/remove small molecules (e.g., water or ammonia to/from a double bond).
- *Ligases* catalyze reactions binding together smaller molecules into larger ones.

Naming Enzymes

The trick to classifying enzymes is to look at the reaction the enzyme catalyzes. Choose a reaction type and apply its proper name. Specific enzyme names are found by finding the substrate (i.e., molecule being acted upon), type of reaction, and adding *-ase*. You can get a lot of information from an enzyme's name.

EXAMPLE 13-8

Different enzymes are closely connected to their substrates. Sucrase catalyzes the hydrolysis of sucrose into glucose and fructose. Lipase catalyzes the hydrolysis of a lipid triglyceride. Alcohol dehydrogenase acts upon an alcohol and pulls hydrogen away (oxidizes it). This also makes it an oxidoreductase.

Krebs Cycle

Hans Krebs, a German-born British biochemist received a Nobel Prize in 1953 for his discovery of the *Krebs cycle* (also known as the citric acid cycle). This physiological cycle has many enzyme-catalyzed chemical reactions that play a key role in oxygen use within all living cells.

In *aerobic* (i.e., oxygen-using) organisms, the citric acid cycle is a major pathway that changes carbohydrates, fats, and proteins into carbon dioxide and water to create a form of usable energy.

Krebs's research into a cell's glucose use and breakdown provided a big leap forward in understanding the metabolism of organisms. Citric acid is the first product and final reactant created at the end of one complete turn of the Krebs cycle. [Note: The Krebs cycle is sometimes called the tricarboxylic acid (TCA) cycle since citric acid is a tricarboxylic acid with three carboxyl groups (COOH).] Figure 13-8 shows the different parts of the Krebs cycle and the enzymes and molecules involved.

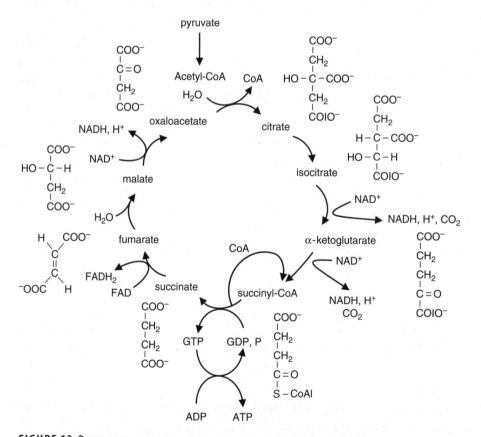

FIGURE 13-8 · The Krebs cycle is loaded with enzyme-catalyzed chemical reactions.

What Are Carbohydrates?

All *carbohydrates* with a carbon, hydrogen, and oxygen in a 1:2:1 ratio, are a part of energy transport, cell surface recognition sites, and essential to DNA and RNA. They get their name from the fact that they are often written as carbon hydrates, $C_n(H_2O)_n$. They are named using the suffix *-ose* (e.g., fructose, $C_6H_{12}O_6$).

> **Carbohydrates** make up a large group of organic compounds and contain carbon, oxygen, and hydrogen.

Carbohydrates make up the largest class of organic compounds in biological organisms. They are products of *photosynthesis*, a reaction between sunlight, carbon dioxide, and the green pigment, chlorophyll.

$$nCO_2 + nH_2O + Energy \rightarrow C_nH_{2n}O_n + nO_2$$

In fact, carbohydrates are a prime source of metabolic energy for plants and plant-eating animals. In addition to nutritional sugars and starches, carbohydrates build and provide strength within cell structures in the form of *cellulose*. Over 50% of the total organic carbon in the Earth's biosphere (living systems) is in the form of cellulose. Plant fibers (e.g., cotton, hemp, flax, and jute) are made up of nearly all cellulose.

With a formula of $(C_6H_{10}O_5)n$, where *n* ranges from 500 to 5000, cellulose is a polymer of glucose. Cellulose's glucose units are linked horizontally like clothespins on a line. (Refer back to Fig. 13-1.) Although single hydrogen bonds are fairly weak, many bonds acting together can provide great stability within large molecules. In fact, this strength makes cellulose indigestible to most mammals, serving instead as digestive roughage.

Sugars

Sugars are a type of carbohydrate. They are more easily converted into energy than any other food. The word *saccharide* (from the Greek word *sákkharon*) means sugar. Saccharides can be different sizes from single sugar molecules called *monosaccharides*, to double (*disaccharides*), triple (*trisaccharides*), and finally huge (*polysaccharides*).

Monosaccharides

Carbohydrates are made up of simple sugar building blocks called monosaccharides. The simple fruit sugars, fructose and glucose, each with the formula $C_6H_{12}O_6$, are monosaccharides.

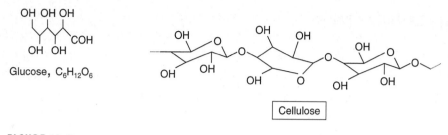

FIGURE 13-9 · Carbohydrates like glucose and cellulose serve many functions in living organisms.

Disaccharides

Disaccharides, or complex sugars, are made up of two monosaccharides. Common disaccharides include table sugar or sucrose (fructose + glucose), and maltose (glucose + glucose). Lactose is a milk disaccharide, formed from one molecule of galactose and one molecule of glucose.

Complex carbohydrates have complicated, folded structures and make up the starch added to plastics, as well as the cellulose in the cell walls of plants and rayon (processed cellulose). They have the formula $(C_6H_{10}O_5)n$, where n is an extremely large number. They are often considered macromolecules because of their huge size compared to simple molecules. Figure 13-9 shows the structure of glucose and cellulose.

Polysaccharides

Polysaccharides, sometimes called *glycans,* are large high-molecular weight molecules created when many monosaccharide units (i.e., up to 500) link up. Polysaccharides are not soluble like mono- and disaccharides. They are too big to be absorbed by the body and must be broken down into smaller units. With highly branched structures, they often have molecular weights of several million. Cellulose, starch, and glycogen are examples of polysaccharides.

Glycosidic Bonds

Polymers are formed from glucose monomers. Two glucose monomers are connected to form a *dimer.* This linkage is made possible by an enzyme catalyst that helps with the removal of a water molecule in a dehydration reaction as shown below.

Enzyme

R-OH + R-OH ➔ R-O-R + HOH

This is a *condensation reaction*, since it condenses two molecules (or R–O–R) into one. When two sugars bond, the –C–O–C– bond is called a *glycosidic bond*. The opposite is true when polymers break down to their constituent monomers. Water is added, and *hydrolysis* takes place. This is shown in the following reaction.

$$R–O–R + HOH \rightarrow R–OH + R–OH$$

Both types of reactions require cell catalysts in order to occur.

Anabolism and Catabolism

When a group of molecules binds to form an even larger molecule, it is known as *anabolism*. During a reaction where a larger molecule is broken down into smaller parts, it is known as *catabolism*. In other words, polysaccharides are broken down into simple monosaccharides through catabolism. In the event that monosaccharides combine to form a complex polysaccharide, it is anabolism or *polymerization*.

? Still Struggling?

Starch (e.g., polysaccharide) can be broken down into glucose by catabolism. When a body needs energy, glucose (e.g., monosaccharide) is pulled from stored polysaccharides through catabolism. If we eat too many starches, glucose is in excess in the body and converted into *glycogen* (e.g., polysaccharide), which is stored in the body by anabolism until needed.

What Are Lipids?

Another group of organic compounds known as *lipids* (e.g., fats, oils, waxes, sterols, and triglycerides) are found in biological systems. They are insoluble in water (i.e., hydrophobic), oily to the touch, and make up another key ingredient within the structure of living cells.

Lipids are similar to carbohydrates in that they bond in groups. Lipid compounds include fats and oils, waxes, steroids, and phospholipids to name a few.

Fats

Fat molecules are compounds (i.e., oils) made up of *glycerol* and three fatty acids bonded in a structure called a *triglyceride*. Glycerol is an alcohol with a hydroxyl group (–OH) on each of its carbons. The structural differences between fats and oils affect their physical chemistry at room temperature. Fats, for example, are commonly solid at room temperature, while oils are liquid. Think of lard (i.e., solid animal fat) compared to olive oil.

Fats provide energy, carry fat-soluble vitamins (A, D, E, K), and are a source of antioxidants. They serve as structural components in brain tissue and cell membranes.

Fatty Acids

Fatty acids are the building blocks of fat molecules. The structure of a fatty acid is made up of a hydrophobic "tail" (i.e., a long hydrocarbon chain) and a hydrophilic "head" of a carboxyl group. Figure 13-10 shows the structure of fatty acids. Fatty acids are used to make soap since their tails are soluble in oily grime, while their hydrophilic heads are soluble in water and help to *emulsify* (see definition in the box on the next page) and wash away dirt.

Saturation of Fatty Acids

In Chap. 12, we learned that hydrocarbons form single and multiple bonds with hydrogen. The number of hydrogen atoms compared to double-bonded carbons

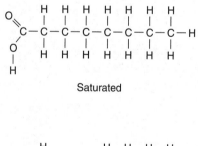

Saturated

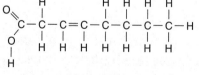

Unsaturated

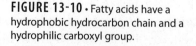

FIGURE 13-10 · Fatty acids have a hydrophobic hydrocarbon chain and a hydrophilic carboxyl group.

in a fatty acid's hydrocarbon tail, makes fatty acids either *saturated, mono-unsaturated,* or *poly-unsaturated.* The carbons in the fatty acid tails of triglyc-erides are all singly bonded to hydrogen atoms. Each carbon with a maximum of four bonds is saturated. These hydrocarbon chains are bunched closely together making them solid at room temperature.

Oils contain double-bonded carbons in the hydrocarbon tail, which allows twists in its molecular shape. Because of these double bonds, carbons can't accommodate four hydrogen atoms and are unsaturated. Bends and twists in the unsaturated hydrocarbon tails keep fats from crowding tightly together. This makes them liquid at room temperature.

Phospholipids

Phospholipids are made up of glycerol, two fatty acids, and a phosphate group. The hydrocarbon tail of the fatty acid is hydrophobic, but the phosphate group at the end of the molecule is hydrophilic since the oxygen has unshared elec-trons. So, phospholipids are soluble in both water and oil.

> An **emulsifying agent** is soluble in oil *and* water, which allows it to mix the two.

As most people know, oil and vinegar don't usually mix. You can whisk them together quickly and get mixing for a few minutes, but then they separate again unless something is added to stabilize the mixture. This is where an emulsifier, such as gelatin or mustard, comes in. An emulsifier has hydrophobic and hydro-philic binding sites, which bind to the soluble and insoluble parts of oil and vinegar.

In living systems, cell membranes with different functions are made up of phospholipids stacked in a double layer with the tails facing each other. The hydrophilic heads face outward on the top and bottom surfaces. Figure 13-11 shows this alignment.

Steroids

If you can spot the central core of four joined rings, you'll have no trouble with *steroids.* This molecular structure is shared by all steroids, including estrogen (estradiol), testosterone, cholesterol, corticosteroids (i.e., cortisone), and vitamin D. Depending on the type of steroid, different groups or whole molecules are attached. Figure 13-12 shows the structure of cholesterol.

Lipoproteins are a combination of lipids and proteins. They transport lipids, such as cholesterol, around in the blood. You may have heard about two main

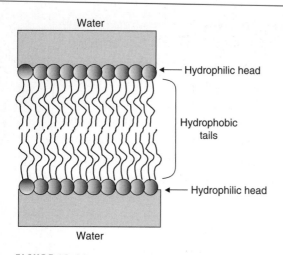

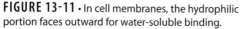

FIGURE 13-11 • In cell membranes, the hydrophilic portion faces outward for water-soluble binding.

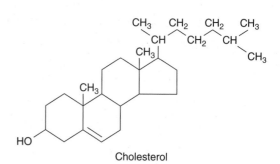

FIGURE 13-12 • Cholesterol, like other steroids, has many attached carbon groups.

lipoprotein types, *LDL* and *HDL*, connected to cholesterol deposition. *Low-density lipoproteins* (LDL) are associated with layering of cholesterol onto artery walls, while *high-density lipoproteins* (HDL) carry cholesterol out of the blood stream to the liver for excretion. As the name implies, HDL has greater density than LDL.

HDL is the "good" cholesterol that removes excess lipids (fat) from the blood. LDL is the "bad" cholesterol that coats lipids onto the inner lining of blood vessels, narrowing them and causing *atherosclerosis*. At the molecular level, HDL has lots of proteins with very little fat, compared to LDL with lots of fat and very few proteins.

Biological Markers

Molecules in the blood, tissues, or other body fluids linked to normal/abnormal processes, conditions, or disease states are called *biological markers* (or *biomarkers* for short).

By using biomarker tests, physicians can find disease, check how well a treatment is working, investigate disease progress, identify toxicity, and create new drugs. In medical use, biomarkers are also called *molecular markers* or *signature molecules*. These molecules are specifically designed to bind to a cell or tissue of medical interest.

A **biomarker** is used to detect the presence of disease, an irregular characteristic, or a physiological condition.

Elevated LDL levels, for example, are a common biomarker of atherosclerosis, while levels of a protein made by the prostate gland, PSA (i.e., prostate-specific antigen), are used to find prostate cancer.

Bioluminescence and Fluorescence

Two common optical biomarker methods are *bioluminescence* and *fluorescence*. Bioluminescence produces light through an enzymatic reaction, while fluorescence comes from light absorption by molecules called *fluorophores*. These molecules have a functional group, which absorbs a specific wavelength of energy and then re-emits it at a different wavelength.

When light is created through a chemical reaction within an organism (e.g., firefly or jellyfish), it is known as **bioluminescence.**

Bioluminescence is commonly used in biology and medicine. Luminescent tags using *luciferase*, an enzyme that emits light when oxidizing the protein luciferin, are as sensitive as radioactive tags. A widely used assay to detect adenosine triphosphate (ATP), which stores energy in a cell, uses firefly luciferase.

Chromophores

Dyes, stains, and pigments, all contain *chromophores*. A chromophore is a functional group capable of specific light absorption that creates color in a substance and binds with other groups to form dyes.

Fluorescent dyes, used to stain tissues and cells for study by fluorescence microscopy, are called **chromophores.**

A chromophore's bonding ability allows it to absorb certain visible light wavelengths, while reflecting other wavelengths. This gives the molecule visible color and, depending on bonding, allows it to emit light. Physicians test for color changes that show different reactions have occurred.

One thing to remember is that any measurement of light can serve as a bio-marker if it provides information on a specific biological state or disease. Length or intensity of fluorescence, as well as light scattering, can also be a biomarker and provide added medical information.

Nanomedicine

Light can be used in different ways. If it hits metal in the body, the metal can get hot enough to cook surrounding tissue (e.g., tumor). If light hits a particle giving off highly reactive oxygen molecules, the oxygen will react with mole-cules in the surrounding tissue and destroy it (i.e., dooming tumors again).

> **Nanomedicine** describes the medical field of targeting disease or repairing dam-aged tissues like bone, muscle, or nerve at the molecular level.

Researchers have applied nano-sized particles of gold-coated glass spheres called *nanoshells* to a process of improving both the detection and treatment of diseased tissue.

> **Nanoshells**, gold-coated silica particles, have tunable optical properties that are affected by size, geometry, and composition.

Nanoparticles made of a silicon core and a gold outer shell have been designed to strongly absorb light wavelengths in the near infrared where light's penetration through tissue is greatest. Figure 13-13 shows the different wave-lengths of the spectrum. (The visible part of the spectrum is further divided according to color, with red at the long wavelength end and violet at the short wavelength end of the spectrum.) In Fig. 13-14, the absorption of different-sized nanoshells is shown.

Cancer therapy uses super small injectable gold nanoshells to travel through a tumor's leaky vessels, target, and then bind to diseased cells. This is called *enhanced permeability and retention*, or EPR effect.

Once injected, gold nanoshells have enough time to move through the tumor vessels into a cancerous tissue (e.g., prostate). Light tuned to the appropriate wavelength heats the gold nanoshell and destroys the cancerous cells. The rising heat (55°C) fries the tumor cells leaving healthy cells unharmed. For surface tumors, nanoshells can be injected directly at the site.

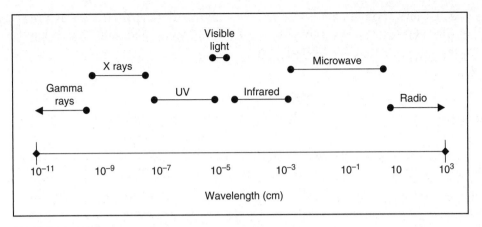

FIGURE 13-13 • There are many different wavelengths of light and energy.

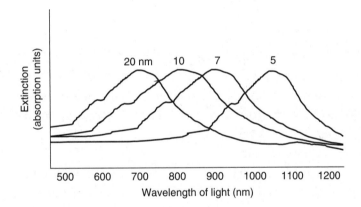

FIGURE 13-14 • Nanoshells can be created or tuned to absorb different wavelengths of light.

The benefit of these targeted nanoshells is that gold nanoshells heat up as they absorb infrared light, but leave healthy tissue (i.e., without bonded nanoparticles) cool and unaffected.

Bioavailability

Since the body has thousands of interconnected subsystems, it is important that medicines are precisely delivered. Our body does this easily, but scientists are just beginning to understand the mechanisms of the body's complex delivery system and how *bioavailability* works.

> **Bioavailability** describes the delivery of drug molecules in the body where they are needed and will do the most good.

Bioavailability and medicine delivery are intricate problems. It's not just a matter of more is better. With toxic chemotherapy drugs, there is a fine line between killing cancer cells and killing the patient.

Medical researchers work at micro- and nanoscales to create new drug-delivery methods, therapeutics, and pharmaceuticals. The diameter of DNA, for example, is in the 2.5 nanometer range, while red blood cells are approximately 2.5 micrometers across. Nanotechnology and nanoscience give physicians powerful new tools in the fight against cancer and degenerative diseases, even aging.

One of the ways nanotechnology works to increase bioavailability is by getting treatment drugs through cell membranes and into cells. Since most virus replication and other disease conditions take place inside a cell, treatments work best within a cell.

Currently, many treatments come to a halt when they get to the cell membrane. (Refer back to Fig. 13-11.) They can't pass through because of charge. Putting polar molecules into a nonpolar membrane doesn't work. One way researchers have gotten around this roadblock is by coating a polar molecule with a nonpolar coating, which is allowed to pass through the membrane and deliver its medical treatment.

Nanomedicine and other medical treatments, based on biochemical innovations, are leading to development of better tools to fight current and future health threats.

QUIZ

1. When polysaccharides break apart into simple monosaccharides, it is known as
    A. polymerization.
    B. anabolism.
    C. stratification.
    D. catabolism.

2. The following are all biological molecules of living cells, organs, and systems, except
    A. proteins.
    B. lipids.
    C. transuranic elements.
    D. nucleic acids.

3. Sucrose, maltose, and lactose are all
    A. monosaccharides.
    B. disaccharides.
    C. polysaccharides.
    D. oligosaccharides.

4. When light is produced via an enzymatic reaction, it is known as
    A. emulsification.
    B. bioluminescence.
    C. fluorescence.
    D. strobe lighting.

5. In the tertiary amine, triisopropylamine, what is the central atom?
    A. Hydrogen
    B. Carbon
    C. Oxygen
    D. Nitrogen

6. Estrogen, testosterone, cholesterol, cortisone, and vitamin D are all examples of
    A. carbohydrates.
    B. proteins.
    C. steroids.
    D. nucleic acids.

7. The following are all carboxylic acids, except
    A. oleic acid.
    B. sulfuric acid.
    C. citric acid.
    D. lactic acid.

8. Which is the largest class of organic compounds in biological systems?
   A. Lipids
   B. Proteins
   C. Esters
   D. Carbohydrates

9. LDL is the "bad" cholesterol that deposits lipids within your blood vessels causing
   A. atherosclerosis.
   B. spina bifida.
   C. osteoporosis.
   D. rickets.

10. A functional group able to absorb specific light wavelengths, create color in a substance, and bind to form dyes is called a
    A. luminore.
    B. biophore.
    C. chromophore.
    D. fluorophore.

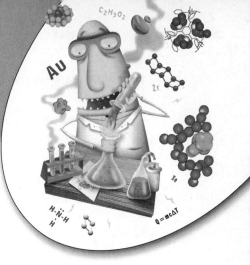

chapter **14**

Environmental Chemistry

In this chapter, you will learn the basics of environmental chemistry, as well as common methods for testing air, soil, and water contamination.

CHAPTER OBJECTIVES

In this chapter, you will

- Learn about the different environmental markers for testing water purity
- Discover the importance of dissolved oxygen
- Find out the difference between point source and non–point source pollutants
- Discover the greenhouse gases involved in global warming
- Understand the chemistry and impacts of acid rain

What Is Environmental Chemistry?

Environmental chemistry includes chemical processes found in nature. It encompasses the source, reactions, movement, and effects of chemicals in the air, soil, and water.

Environmental chemists work with specific units, solubility, and chemical reactions to further understand what is happening to different environmental chemicals.

Contamination

Environmental *contaminants* are found at higher than normal levels in nature or in places they don't belong. Similar to *pollutants*, which have harmful impacts on surroundings, contaminants may seem harmless at first, but can cause toxic or harmful effects over time.

> A material or organism affected by a pollutant is called a **receptor**, while a **sink** is a material or species that stores and interacts with a pollutant (e.g., mercury).

Four main contaminant types exist: organic, inorganic, acid/base, and radioactive. They are released into the environment in many ways, but most chemicals enter the hydrologic cycle (i.e., water cycle) as direct (*point source*) or indirect (*non–point source*) contamination.

Point source contamination (e.g., bilge water, diesel leaks, factory sewage, refinery oil, and waste-treatment plants) is released directly into urban water supplies. In the United States and elsewhere, these releases are monitored, but some contaminants find their way into native waters. Table 14-1 lists some of the toxic contaminants found in soil and water.

TABLE 14-1 Many toxic contaminants are found in soil and water.	
Compound	**Toxic Levels (ppm)**
Arsenic (playground soil)	10.0
Arsenic (mine tailings—toxic)	1320
Diethyl ether	400
Trihalomethane (water)	0.10
Nitrate (water)	10.0
Nitrite (water)	1.0
Silver (water)	0.05
Cadmium (water)	0.005
Mercury (water)	0.002

Non–point source contamination (e.g., storm water and agricultural runoff) enters the water supply from soils/groundwater runoff and the atmosphere through rainfall. Soils and groundwater take in fertilizer and pesticide residues as well as industrial waste. Atmospheric contaminants come from the gaseous emissions of automobiles, factories, and even restaurants.

Environmental Markers

Environmental chemists test water purity by several methods. These tests include checking environmental levels for the following:

- Dissolved oxygen (DO)
- Chemical oxygen demand (COD)
- Biological oxygen demand (BOD)
- Total dissolved solids (TDS)
- pH—acid or base levels of a sample
- Nitrates, phosphorus, and pesticides
- Heavy metals (e.g., copper, zinc, cadmium, lead, and mercury)
- Radioactivity (more on radioactive contamination in Chap. 15)

> **Dissolved oxygen** is the amount of oxygen measured in a stream, river, or lake.

Biological Oxygen Demand

Biological oxygen demand (BOD) is a chemical test for finding the amount of dissolved oxygen needed by marine organisms to break down organic material in a given water sample at a set temperature over a specific time.

Fish pull oxygen from water and absorb dissolved oxygen through their gills. So, dissolved oxygen is an important marker of a river or lake's capacity to support marine life. The actual level of dissolved oxygen present in even the cleanest water is extremely small (e.g., <1 mg/L).

> **Biological oxygen demand** measures the oxygen used by microorganisms to decompose organic waste.

The amount of dissolved oxygen in water depends on several factors, including temperature (the colder the water, the more oxygen that can be dissolved), volume and velocity of water flow, and number of organisms using oxygen for respiration.

? Still Struggling

Oxygen solubility in water at a temperature of 20°C is 9.2 milligrams oxygen per liter of water (i.e., around 9 parts per million). The concentration of oxygen dissolved in water is written in milligrams per liter (mg/L) of water.

Chemical Oxygen Demand

A *chemical oxygen demand* (COD) test is often used to check the amount of organic compounds in water.

The COD test is based on the fact that nearly all organic compounds are fully oxidized to carbon dioxide with a strong oxidizing agent under acidic conditions. The amount of oxygen needed to oxidize an organic compound into carbon dioxide, ammonia, and water is given by:

$$C_nH_aO_bN_c + \left(n + \frac{a}{4} - \frac{b}{2} - \frac{3}{4}c\right)O_2 \rightarrow nCO_2 + \left(\frac{a}{2} - \frac{3}{2}c\right)H_2O + cNH_3$$

This equation does not include the oxygen demand caused by the oxidation of ammonia into nitrate. When ammonia is converted into nitrate, it is known as *nitrification*. Ammonia's oxidation into nitrate is shown in the following equation:

$$NH_3 + 2O_2 \rightarrow NO_3^- + H_3O^+$$

The second part of the equation includes oxidation from nitrification.

Turbidity

Turbidity measures the murkiness (i.e., ability to see through it) of a solution. The murkier a solution, the higher its turbidity. Water turbidity is caused by suspended material, such as clay, silt, and organic matter. Plankton and other microorganisms also block the path of light through water. A *turbidity current*, caused by underwater sliding of deep ocean sediments, piles a dense slurry on the ocean floor like a mud slide.

Environmental pH

As explained in Chap. 10, a solution with a pH of 7 is neutral, while a solution with a pH of less than 7 is acidic and a solution with a pH of greater than 7 is basic. Natural water reservoirs commonly have a pH between 6 and 9. The pH scale is negatively logarithmic, so each whole number (reading downward) is 10 times the preceding one (e.g., pH 5.5 is 100 times as acidic as pH 7.5).

$$pH = -\log [H+] = \text{hydrogen ion concentration}$$

The pH of natural waters becomes acidic or basic as a result of human activities such as acid mine drainage, emissions from coal-burning power plants, and heavy automobile traffic.

What Is Acid Rain?

As long ago as the seventeenth century, industry and acidic pollution were known to affect plants and people. Robert Angus Smith, a Scottish chemist, was interested in environmental issues such as public health, disinfection, and peat formation at this time.

Famous for research on air pollution, Smith described *acid rain* in 1872, and published a book called *Air and Rain*. Smith's work was also noticed by England's Queen Victoria who appointed him as the first inspector under the British Alkali Acts Administration of 1863. Smith called his atmospheric work chemical climatology.

Although this first awareness of environmental contamination was important, it took another 100 years before acid rain problems became internationally known.

> **Acid rain** includes all types of precipitation (rain, snow, sleet, hail, fog) that are acidic (pH lower than the 5.6 average of rainwater) in nature.

Rain forms carbonic acid naturally when it reacts with carbon dioxide in the atmosphere as shown in the following reaction:

$$CO_2 + H_2O \rightarrow H_2CO_3$$

Precipitation, normally between 5.0 and 5.6 in pH, is slightly acidic. However, some sites on the eastern North American coast have precipitation pH levels near 2.3 or roughly 1000 times more acidic than pure water because of various additional pollutants. (Note: Vinegar, which is very acidic, has a pH of 3.0.)

> **Dissolution** takes place when water molecules, acids, or other environmental compounds attract and remove oppositely charged ions from rock.

Historical treasures such as statutes and buildings, hundreds to thousands of years old, suffer chemical weathering through a process called *dissolution*. Limestone ($CaCO_3$) dissolution by acid rain (H_2SO_4) takes place by:

$$CaCo_3 + H_2SO_4 \rightarrow CaSO_4 + H_2O + CO_2$$

Sulfur and nitrogen oxides forming acid rain belch from industrial smokestacks, vehicle exhausts, and burning wood. In the atmosphere, they mix with water molecules in the clouds and change to sulfuric and nitric acid. Then, rain and snow wash these acids out of the air.

? Still Struggling

Acid rain comes mostly from three atmospheric compounds (i.e., CO_2, NO, and SO_2) found in the atmosphere that react with water and fall to the earth in acidic raindrops. When acid rain falls onto limestone statues, monuments, and gravestones, it dissolves, discolors, and/or ruins the surface by reacting with the rock's elements.

Nutrient Leaching

Acid rain impacts forests and plant life resulting in nutrient leaching and the concentration of toxic metals. *Nutrient leaching* happens as acid rain adds hydrogen ions to the soil, which in turn reacts with existing minerals. These interactions strip calcium, magnesium, and potassium from soil and rob trees of nutrients.

Toxic metals (e.g., lead, zinc, copper, chromium, and aluminum) are deposited in the forest by the atmosphere. When acid rain interacts with these metals, tree and plant growth are stunted along with mosses, algae, nitrogen-fixing bacteria, and fungi. When mercury and aluminum are released into the environment through acidification of the soil, they can leach into groundwater.

High metal levels are toxic to humans. Aluminum has been linked to Alzheimer's disease. High concentrations of sulfur dioxide and nitrogen oxides have been linked to increased respiratory illness in children and the elderly.

Acid deposition also affects jobs. Acidic lakes and streams negatively affect fish populations. Lower fish numbers impact commercial fishing and industries dependent on sport fishing revenues. Forestry and agriculture are also affected by crop damage.

Greenhouse Effect

Greenhouse gases trap the sun's warmth and maintain the Earth's surface temperature at a level needed to support life. Some heat is reflected back into space, but a lot of it is locked in the atmosphere by greenhouse gases, which acts like a lens and heats the Earth. Figure 14-1 shows how these greenhouse gases act in the atmosphere to trap energy.

> The **greenhouse effect** describes how atmospheric gases prevent heat from being released back into space, allowing it to build up in the Earth's atmosphere.

If the global greenhouse effect gets too strong, it can heat the Earth faster than biological organisms can adapt. The problem is that even a small amount of heat (i.e., a few degrees more) creates problems for people, plants, and animals.

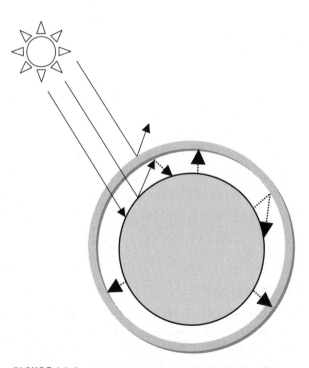

FIGURE 14-1 • Greenhouse gases trap incoming heat from the sun causing the planet to heat.

Global Warming

Greenhouse gases are a natural part of the atmosphere and include water vapor, carbon dioxide, methane, nitrous oxide, fluorocarbons, and ozone.

Global warming comes from increases in carbon dioxide, methane, and nitrous oxide amounts. The planet is warming at a steep rate. This warming has been predicted by climate models using many environmental variables (e.g., gases released yearly into the atmosphere). The consequences of global warming, such as the melting of the polar ice sheets, rising sea levels, and species' range and migration changes have made news worldwide.

Carbon Dioxide

Carbon dioxide (CO_2) is a natural greenhouse gas and also the biggest human-supplied gas to the greenhouse effect (about 70%). A heavy, colorless gas, carbon dioxide is the main gas we exhale during breathing. It dissolves in water to form carbonic acid, is formed in animal respiration, and comes from the decay or combustion of plant/animal matter. Carbon dioxide is also pulled from the air by plants during photosynthesis and used to carbonate drinks.

Unfortunately, the Earth's inhabitants don't have the option to stop breathing. However, the amount of carbon dioxide in the atmosphere is currently about 30% higher now than it was at the beginning of the 1800s. Figure 14-2 shows carbon dioxide concentration trends of past centuries.

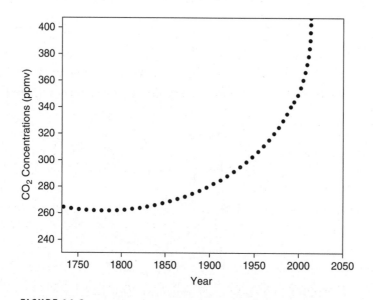

FIGURE 14-2 · Carbon dioxide trends of the past 250 years.

TABLE 14-2 A variety of gases are released during a volcanic eruption.

Volcanic Gas	Percentage of Total Gases (average)
Water vapor (steam) and carbon dioxide	90–95
Sulfur dioxide	< 1
Nitrogen	< 1
Hydrogen	< 1
Carbon monoxide	< 1
Sulfur	< 0.5
Chlorine	< 0.2

The industrial revolution is responsible for some of this jump. Ever since fossil fuels such as oil, coal, and natural gas were first burned for electricity and fuel, carbon dioxide levels began to climb. Additionally, when farmers and home owners clear and burn weeds and crop stubble, carbon dioxide is produced.

Carbon dioxide gases come from the Earth as well. When volcanoes explode, about 90–95% of the spewed gases are made of water vapor and carbon dioxide. Table 14-2 lists the different gases released into the atmosphere during a volcanic eruption.

Nature attempts to balance the carbon in the Earth's systems by storing it away in different forms. This is an area of intense environmental research since scientists don't know how long various carbon storage mechanisms (e.g., oceans) will continue to take up increased carbon before reaching their limits.

Nitrogen Oxide

The colorless gas known as nitrous oxide is an atmospheric pollutant produced by combustion. As a greenhouse gas, nitrogen oxides trap heat much more efficiently than carbon dioxide.

There are several ways nitrogen and oxygen team up in the atmosphere, including nitric oxide (NO), nitric acid (HNO_3), and nitrous oxide (NO_2). Nitrogen oxides are stable gases and don't break down quickly. For this reason, they build up in the atmosphere in greater and greater concentrations. Nitrogen dioxide in the sky causes the yellow-brown haze many people call *smog*. Figure 14-3 shows how the nitrogen cycle works.

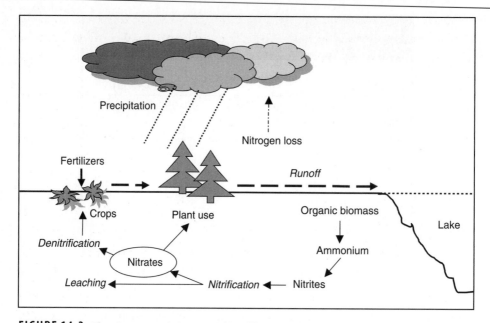

FIGURE 14-3 · The nitrogen cycle is essential to all living systems.

Nitrogen combines with atmospheric water to form nitric acid. This comes down as rain and acidifies lakes and soils, killing fish and small animal populations, and damaging forests. Acid particulates are also washed out, along with the leaching of heavy metals, into water supplies. Scientists believe that increased nitrogen pollution comes from crop burning, industrial releases, and fertilizers used in agriculture.

Nitrous oxide, called *laughing gas*, is used by dentists to put patients to sleep during dental procedures. The amount of nitrous oxide in the atmosphere is about 15% higher now than it was in the 1800s.

Besides the burning of fossil fuels, nitrogen oxides are also produced by kerosene heaters, gas ranges and ovens, incinerators, deforestation and leaf burning, as well as aircraft engines and cigarettes. Lightning and soil sources also produce nitrogen oxides. Scientists estimate that vehicles produce 40% and electric utilities and factories produce 50% of industrial nitrogen oxide emissions. The remaining 10% comes from other sources.

Methane

Another greenhouse gas, *methane*, is a colorless, odorless, flammable hydrocarbon released by the decay of organic matter and formation of coal. This gas is the second biggest addition, after carbon dioxide, to the greenhouse effect at around 20%.

Methane, the main component of natural gas, is found in deposits, (e.g., oil) in the Earth's crust. Methane is a by-product of the production, transportation, and use of natural gas. Plants decaying underwater create methane sometimes called *marsh* or *swamp gas.*

Farm animals are one of the best known sources of methane in rural areas. Belching cows, with complicated digestive systems, release huge amounts of methane in satisfying burps. It sounds funny, but when you consider herds of hundreds or thousands of animals, it adds up!

The U.S. National Aeronautics and Space Administration (NASA) reports that methane levels are rising in the atmosphere three times faster than carbon dioxide; atmospheric methane levels have tripled in the past 30 years. Over 500 million tons of methane are emitted yearly from bacterial decomposition and fossil fuel burning.

The amount of atmospheric methane is about 145% higher now than it was in the 1800s. The major causes of this increase are thought to include

- Digestive gases of sheep and cattle
- Growth and cultivation of rice
- Geologic release of natural gas
- Decomposition of garbage and landfill waste

Methane's interaction in the atmosphere is fairly complex. Naturally occurring hydroxyl radicals in the atmosphere combine with and pull methane from the atmospheric mix. Unfortunately, methane and carbon monoxide (from car emissions) lower hydroxyl levels; so, as hydroxyl concentrations drop and methane emissions continue, methane concentrations rise.

Nature's Balancing Act

Green plants use the sun's energy and carbon dioxide from the air as part of photosynthesis. This is a good thing since they soak up carbon dioxide in the process (i.e., carbon sequestration). Figure 14-4 shows the atmospheric carbon cycle.

During a photosynthesis cycle, plants form carbohydrates; key components of the food web. Plants also serve as carbon dioxide storehouses. Maybe this is why some people think talking to houseplants is good for them. The carbon dioxide exhaled during breathing is a critical part of a plant's energy process. The more you talk to them, the more carbon dioxide you provide.

Forests absorb carbon dioxide in a big way. Over time, forests build up a significant supply of stored carbon in their tree trunks, roots, stems, and leaves.

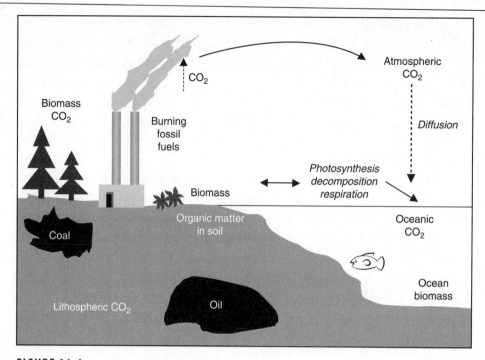

FIGURE 14-4 · In modern times, the carbon cycle has a large industrial component.

Then, when the land is cleared, this stored carbon is converted back to carbon dioxide by burning or decomposition.

The oceans are another big player in carbon dioxide absorption. They pull carbon dioxide from the atmosphere and create a balancing influence on temperature ranges. Table 14-3 lists estimated carbon amounts stored in various areas.

TABLE 14-3 Carbon is stored in several natural repositories.

Carbon Storage	Quantity (billions of metric tons)
Atmosphere	580 (1700) 800 (2000)
Organic (soil)	1500–1600
Ocean	38,000–40,000
Ocean sediments and sedimentary rocks	66,000,000–100,000,000
Land plants	540–610
Fossil fuels	4000

Ozone

In 1995, three chemists, Mario Molina, Sherman Rowland, and Paul Crutzen warned world leaders of damage being done to the *ozone* (O_3) layer. This natural layer of ozone molecules, located from 9 to 30 miles up in the atmosphere, blocks the Earth from the sun's damaging ultraviolet radiation. The chemists discovered human-made compounds [e.g., chlorofluorocarbons (CFC) and nitrogen oxides] used as refrigerants and propellants in spray cans, reacted with atmospheric ozone and reduced it. For their work, they received the 1995 Nobel Prize in Chemistry.

In response to the ozone depletion problem, chemists started experimenting with refrigerants that didn't involve ozone. Substitutes were found and the environmental impact reduced.

Many elements used today were discovered by using cutting-edge technology and equipment. Since the 1960s, most elements added to the Periodic Table were human-made and not naturally occurring. These molecules have complex uses that many research and applications chemists and biochemists are just beginning to understand.

Biodegradable

Chemists working in the plastics industry were heavily criticized when landfills became overloaded with disposable plastic containers and a softer compound called Styrofoam. Environmentalists sounded the alarm for consumers to think twice before buying products, especially fast food, packaged in these synthetic containers.

> Any material, which can be broken down into simpler components by microorganisms (e.g., bacteria) is **biodegradable.**

In order to meet the new concern, chemists doubled their interest in the biodegradability of plastic products. They discovered the addition of complex carbohydrates (polysaccharides) to plastics allowed microorganisms to break down the plastic products. Now several companies, restaurants, and natural groceries stores (e.g., Whole Foods Market™) are using biodegradable utensils, plates, and take-out boxes/bags.

Tough Issues

The environment is a very complex mixture of elements. Industrialized cities have much higher levels of metals and acids in their air than rural areas. Biochemical engineers and environmental chemists must work together to test air and water samples, as well as industrial emissions in their efforts to piece together the overall environmental impacts of modern life. The interconnectedness of all life forms also affects the necessity of controlling environmental pollution.

QUIZ

1. Wastewater, diesel leaks, and sewage released from factories, refineries, and waste-treatment plants directly into water supplies is an example of
   A. green chemistry.
   B. non–point source contamination.
   C. point source contamination.
   D. dissolution.

2. The pH scale is negatively logarithmic and based on the concentration of which ion?
   A. Iodine
   B. Potassium
   C. Sodium
   D. Hydrogen

3. Which natural greenhouse gas contributes the most to the greenhouse effect?
   A. Methane
   B. Carbon dioxide
   C. Nitrogen oxide
   D. Propane

4. What was added to plastics products that allowed microorganisms to break them down?
   A. Polysaccharides
   B. Lipids
   C. Steroids
   D. Esters

5. Which of the following methods is used to test the quantity of organic compounds in water?
   A. Turbidity test
   B. Chemical oxygen demand test
   C. Radioactivity detector
   D. Heavy metal

6. What serves as a lens to trap the sun's heat and reactive greenhouse gases, raising the Earth's surface temperature?
   A. Forests
   B. Oceans
   C. Cosmic dust
   D. Atmosphere

7. Which of the following is an important indicator of a lake's ability to support marine life?

A. Carbon tetrafluoride

B. Free nitrogen

C. Dissolved oxygen

D. Not many fishermen

8. Any material broken down into simpler components by microorganisms is considered

A. hazardous.

B. inorganic.

C. biodegradable.

D. insoluble.

9. Which colorless, odorless, flammable hydrocarbon released by the breakdown of organic matter and coal carbonization coal is known also as swamp gas?

A. Propane

B. Ethane

C. Pentane

D. Methane

10. What is the name of the process that causes the surface of limestone statues to be discolored and disfigured by acid rain?

A. Glaciation

B. Dissolution

C. Desertification

D. Sedimentation

chapter 15

Nuclear Chemistry

Nuclear chemistry can bring up terrible historical images; huge mushroom clouds over Nagasaki and Hiroshima in 1945 after the detonation of atomic bombs. Several thousand people were instantly vaporized (i.e., at ground zero) or felled later by radiation sickness. Information from the former USSR about the Chernobyl disaster in 1986 revealed how a nuclear reactor explosion had released radioactive gases and contaminated the atmosphere over Eastern Europe.

But the news is not all bad. Nuclear chemistry also provides powerful tools to diagnose health problems and save lives. Through nuclear imaging (e.g., PET scans) and radioactive treatments, cancer cells can be pinpointed and eliminated. Nevertheless, these examples are all human applications of a natural process. It took many scientists working long hours to understand nuclear chemistry and develop potential applications, good and bad.

In this chapter, you will learn how radioactive isotopes, half-life, emission particles, and radioactive markers are used in medicine and other applications.

CHAPTER OBJECTIVES

In this chapter, you will

- Understand reactions involving the nucleus
- Learn about radioactivity and its effects
- Become familiar with magic numbers and atomic nuclei
- Understand half-life and radioactive decay

What Is Radioactivity?

In 1896, Antoine Becquerel, a French physicist, discovered radioactivity of chemical elements, sometimes called *radiochemistry*. He noticed a photographic plate in his lab had become exposed by sunlight. The only possible explanation was a uranium salt sample sitting nearby on the bench top. This was puzzling. How could a solid compound cause a chemical reaction through the air without any contact?

The term, *radioactivity*, was first used by French scientist, Marie Curie in 1898. Marie and her physicist husband, Pierre, found there were two kinds of radioactive particles emitted from compounds. They called the electrically negative (−) kind beta (β) particles and positive (+) kind alpha (α) particles.

These early years of discovery illustrated the special properties of radioactive elements and led to more studies by Pierre and Marie. In 1903, Becquerel and the Curies shared the Nobel Prize in Physics for their work in radioactivity. In 1911, after the discovery of polonium and radium, the Curies received another Nobel Prize, this time in chemistry for their extensive work.

Isotopes

In 1912, research into the atomic mass of different elements ran into problems. Some elements seemed to have forms with different masses, but the same chemical properties. When they were from the same element, they had the same atomic number or number of protons. The mass differences were connected to their number of neutrons. The elements weren't new, just different forms of previously identified elements.

In 1913, Frederick Soddy, a British chemist, called chemically identical elements with different atomic weights, *isotopes*, from the Greek word meaning "same place." These were placed in the same spot on the Periodic Table.

> **Isotopes** are different types of atoms with the same number of protons (i.e., atomic number), but different numbers of neutrons (i.e., atomic mass).

Three Forms of Hydrogen

Even though it is the simplest element, hydrogen actually exists in three different forms as shown in Fig. 15-1. The most common form of hydrogen is called *protium*. It has only one proton and a mass of one. In 1931, Harold Urey, an

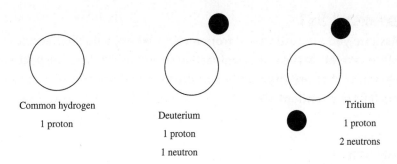

FIGURE 15-1 • Hydrogen has three different forms that are named separately.

American physical chemist, was isolating hydrogen when he found a form with twice the mass of common hydrogen. He named this hydrogen isotope *deuterium* after the Greek word for two, *deuteros*.

In nature, there is 1 deuterium for every 5000 regular hydrogen atoms. Urey also proved water could be formed with two deuterium atoms instead of hydrogen atoms to make "heavy water." Based on this work and other studies with isotopes, he earned the Nobel Prize for Chemistry in 1934. An even rarer radioactive isotope of hydrogen, *tritium*, has a mass of three, with one proton and two neutrons.

Naming Isotopes

Hydrogen is the only element with separate names for its different isotopes. For the isotopes of other elements, two shorthand naming methods are commonly used. One is to write the element name with a hyphen and then the atomic mass. The second method uses the element symbol along with its atomic number (Z) as a subscript *and* the atomic mass as a superscript. Both are written to the left side of the element symbol. (Note: Atomic mass is not the same as atomic number!)

Figure 15-2 shows both ways of writing isotopes for radioactive radon, which has over 20 different isotopes. Radon-222 is the longest-lived of these isotopes with a half-life of nearly 4 days.

$$\text{Rn-222}$$

$$^{222}_{86}\text{Rn}$$

FIGURE 15-2 • There are two different ways to name radioactive isotopes. Radioactive radon is shown.

EXAMPLE 15-1

Assume you are testing soil from an area where a nuclear accident took place several years earlier. Radioactive strontium ($^{90}_{38}$Sr) is suspected to be present. What are the number of protons and neutrons in the isotope nucleus of strontium-90?

SOLUTION

First, the atomic number (i.e., number of protons) of (Sr) is written as a subscript to the *lower* left of the element symbol. The mass number (protons + neutrons) is written as a superscript in the *upper* left of the element symbol.

You can find the number of neutrons by subtracting the number of protons from the atomic mass (the upper-left superscript) as shown in the following example:

$$\text{Number of neutrons} = 90$$
$$\text{Number of neutrons} = 38$$
$$90 - 38 = 52$$

Radioactive strontium isotope has 38 protons and 52 neutrons ($^{90}_{38}$Sr)

EXAMPLE 15-2

What are the atomic numbers for Sr-90, Cl-37, and Mg-24?

SOLUTION

Did you get 38, 17, and 12?

EXAMPLE 15-3

Can you name these unknown elements $^{108}_{47}$X, $^{238}_{92}$X, and $^{98}_{43}$X?

SOLUTION

Did you get silver, uranium, and technetium?

Nuclear Reactions and Balance

Most chemical reactions involve an element's outer electrons as they are shared, swapped, and bumped. Nuclear reactions are different. All the action takes place inside the nucleus.

There are two types of nuclear reactions. The first is the radioactive decay of bonds within the nucleus that emit radiation when broken. The second is the "billiard ball" type of reaction where a nucleus or a nuclear particle (like a proton) collides with another nucleus or nuclear particle.

What Is Radioactive Decay?

Even though elements exist as different isotopes, they are not always stable. Radioactive decay takes place when a nucleus reorganizes itself to become more stable. Figure 15-3 demonstrates the way radioactive elements decay. This process gives off energy in the form of ionizing particles or radiation. No collision with other atoms is needed, it just happens spontaneously. The original atom is called the *parent nuclide*, while the atom after the emission is called the *daughter nuclide*.

> **Radioactive decay** takes place when an unstable atomic nucleus loses energy by emitting ionizing particles or radiation.

For example, carbon-14 (i.e., parent nuclide) emits radiation to become nitrogen-14 (i.e., daughter nuclide). While all radioactive isotopes go through this process, each one has its own *decay rate*.

Element→Decay →Radiation

1.	*Element* ➔ decay ➔ radiation ➔ *electrons, neutrons, smaller nuclei, and electromagnetic radiation.*
2.	<u>*Element A*</u> ➔ <u>*Element B*</u> ➔ <u>*Element C*</u> (Different form) Nucleus 1 Nucleus 2

FIGURE 15-3 · Radioactive elements decay and lose energy in an ordered way.

Different elements have very different decay rates. The decay rate of an element's radioactive isotopes is affected the element's stability at a certain energy level. Bismuth (Bi), at atomic number 83, is the heaviest element in the Periodic Table with a minimum of one stable isotope. All other heavier elements have radioactive isotopes.

Transuranium Elements

Uranium is the element with the highest atomic number found in nature; all elements with greater atomic numbers are called *transuranium elements*. The transuranium elements of the actinide series were discovered as synthetic radioactive isotopes at the University of California at Berkeley or the Argonne National Laboratory. By colliding uranium (U-238) with neutrons, uranium (U-239) was produced which decayed days later to neptunium (Np-237). Other experiments led to the creation of americium (Am-243) and curium (Cm-247).

Radioactive elements after uranium in the Periodic Table (i.e., $Z > 92$), with the exception of Pu-244 and Pu-239, are all artificially made through nuclear reactions. This is one reason actinide series elements have similar chemical properties. Table 15-1 gives examples of transuranic elements.

Radiation Emission Types

Three main radiation types are given off alone or in combination during the breakdown of radioactive elements. They are *alpha* (α) particles, *beta* (β) particles, and *gamma* (γ) rays. Alpha and beta particles are dangerous to living things since they penetrate cells and damage proteins. However, gamma rays are much more penetrating and harmful, stopped only by thick, dense metals like lead. Gamma rays are high-energy electromagnetic waves like light, but with a shorter, more penetrating wavelength.

A radioactive element, like everything else in life, decays (ages). When uranium or plutonium decays over billions of years, it goes through a transformation process of degrading into lower-energy element forms until it settles into a stable configuration.

As mentioned above, when a radioactive element decays, different nuclear particles are given off. These radioactive particles can be separated by an electric (magnetic field) and detected in the laboratory:

Alpha (α) particles = positively (+) charged particles

Beta (β) particles = negatively (−) charged particles

Gamma (γ) particles = high-energy particles with zero charge

TABLE 15-1	Transuranic elements are human-made via nuclear reactions	
Element	**Symbol**	**Atomic Number**
Neptunium	Np	93
Plutonium	Pu	94
Americium	Am	95
Curium	Cm	96
Berkelium	Bk	97
Californium	Cf	98
Einsteinium	Es	99
Fermium	Fm	100
Mendelevium	Md	101
Nobelium	No	102
Lawrencium	Lw	103
Rutherfordium	Rf	104
Dubnium	Db	105
Seaborgium	Sb	106
Bohrium	Bh	107
Hassium	Hs	108
Meitnerium	Mt	109
Darmstadtium	Ds	110
Roentgenium	Rg	111
Not named	—	112, 114, 115

Alpha Particles

Alpha decay occurs when a nucleus has too many protons and excessive repulsion. Alpha particles are commonly given off or *emitted* by larger radioactive isotopes (e.g., uranium, thorium, actinium, and radium, and transuranium elements).

Alpha particles are emitted as radiation and made up of two protons and two neutrons. Without electrons, they have two positive charges. As charged particles, alpha particles can be trapped by a magnetic field, which helps scientists identify them. But once released, alpha particles usually pick up electrons to eventually become helium atoms.

The loss of an alpha particle decreases the atomic mass of the parent (i.e., original) nuclide by four and the atomic number by two. In other words, the element is transformed. Alchemists tried for centuries to change common elements, like lead, into gold without success. During radioactive decay, nature already has a way to change elements from one form to another. But, you still can't make gold through radioactive decay.

Although alpha particles are the strongest ionizing forms of radiation, they are usually not harmful biologically unless eaten or inhaled. Alpha particles do not penetrate very deeply and your skin usually stops them.

EXAMPLE 15-4

When an isotope such as radon-222 decays, the alpha emission equation looks like the following:

$$^{222}_{86}Rn \rightarrow {}^{218}_{84}Po + {}^{4}_{2}He$$

When radon-222 decays to polonium, the atomic number is reduced by 2 and the atomic mass by 4.

EXAMPLE 15-5

The same thing happens when uranium-238 decays to thorium and helium:

$$^{238}_{92}U \rightarrow {}^{234}_{90}Th + {}^{42}He$$

Beta Particles

Beta particles, emitted as radiation, are high-energy, high-speed electrons (−) or positrons (+) that come from the nucleus. When atoms like potassium-40 give off beta particles, there is a net charge of zero. This process happens in three different ways:

1. Neutron-proton ratio ↑, so a neutron (neutral) changes into a proton (+) and releases an electron (− beta particle).
2. Neutron-proton ratio ↓, so a proton (+) changes into a neutron (neutral) and releases a positron (+ beta particle).
3. Neutron-proton ratio in the nucleus ↓, so the nucleus captures an electron (−) and changes a proton (+) into a neutron (neutral).

Beta particles have average ionizing and penetrating power. They are between the strength of alpha particles and gamma rays. Beta particles, particularly strontium-90, are used for treating eye and bone cancers. Positron or beta plus (β+) decay, can be helpful as a tracer during *positron emission tomography*, as in a PET scan. Beta particles are also used to find the paper thicknesses in quality control tests.

Positron emission tomography (PET), a nuclear medicine imaging technique, provides three-dimensional (3D) images of functional processes in the body. PET imaging detects pairs of gamma rays emitted indirectly by a *positron-emitting radionuclide* (tracer), introduced into the body on a biologically active molecule. Tracer concentration images in 3D or 4D space (i.e., over time) within the body are then reconstructed by computer modeling. (More about nuclear medicine later in the chapter.)

? Still Struggling

Electron capture takes place in an energy level (1s) near the nucleus. After a proton grabs an electron from a previously filled (↑↓) energy level, there is an empty slot (↑_) in the (1s) energy level. Since nature abhors a vacuum, an electron from a higher energy level jumps down to fill the vacant slot, leaving a vacancy at its higher level. To fill that empty slot, an electron from an even higher energy level jumps down. Subsequently, a domino effect of downward electron jumps takes place and energy is released (i.e., x-ray part of the electromagnetic spectrum) making it easy for scientists to measure the released energy.

Gamma Rays

Gamma rays are high-energy, short-wavelength photons. Unlike alpha and beta particles, they have no mass or charge. Gamma decay takes place when energy in the nucleus is too high and must drop to a lower energy state. When this happens, a gamma ray is emitted.

Emitted gamma rays are an extreme health hazard since they are one of the most penetrating forms of radiation, even though the weakest. Different from alpha and beta particles, which usually burn the skin's surface, gamma rays sink into tissue and damage cells.

Radiation Detection

As in other parts of chemistry, there are International Standard (SI) units and common units for radiation measurements. The amount of radiation given off by a radioactive compound is measured using the *curie* (Ci), named for the famed scientist, Marie Curie, or the SI unit *becquerel* (Bq). The Ci or Bq express the number of radioactive atom disintegrations in a radioactive compound over time.

For example, one Ci is equal to 37 billion (37×10^9) disintegrations per second. Similarly, 1 Bq is equal to 1 disintegration per second, so that one Ci is equal to 37 billion (37×10^9) Bq. These units are often used to describe the amount of radioactive materials released into the environment.

On April 26, 1986, the Chernobyl nuclear power plant in the former Soviet Union (i.e., Ukraine) was the site of the worst nuclear accident in history. An estimated 81 million Ci of radioactive cesium was released into the surrounding environment. By comparison, the partial core meltdown at the Three Mile Island Nuclear Generating Station on March 28, 1979, in the United States released 13 million Ci of radioactive gases, but less than 20 Ci of the especially unsafe iodine-131.

Types of Detectors

Although just curiosities at first, radioactive elements have become important tools in basic research and medicine. However, nuclear scientists needed a way to detect radioactive elements and any lingering radioactivity. This was solved with the development of radiation particle counters. Today, these tools are critical for safely monitoring radioactivity and performing scientific experiments.

In 1928, H. Geiger and E. W. Muller invented the *Geiger counter* (also known as Geiger-Muller). Their device counted the particles emitted by radioactive nuclei, which collided with atoms of a nonreactive noble gas (e.g., argon) that had been sealed in a tube. The collisions, which caused atoms to briefly ionize and conduct electricity, made a needle move on a detector.

A Geiger counter can also be set to make noises like the familiar beeps you may have heard in science fiction movies. The higher the radioactivity levels, the faster and more beeps are heard. Geiger counters can detect beta particles and gamma rays, as well as some alpha particles. These radiation detectors are sturdy and cheap to make, so they are used all the time, especially in laboratories. Safety first!

Similarly, Sir Samuel Curran invented the *scintillation counter* in 1944 for measuring ionizing radiation as part of the Manhattan Project (i.e., building the atomic bomb). His device was based on Becquerel's earlier radioactivity work. It counted the number of flashes produced to gain an estimate of the radioactivity of the source, that is the more flashes of light seen the more radiation present.

A scintillation counter uses a scintillator or crystal sensor, (e.g., phosphor) to detect tiny flashes of light produced by the phosphor when it absorbs ionizing radiation. When they collide, the chemical excites and emits fluorescent light. Scientists count the number of light flashes to get an estimate of a sample's radioactivity level. Scintillation counters are more sensitive than Geiger counters, but they are big machines not easily moved. Any samples to be tested must be taken to the equipment. Scintillation counters are used in research laboratories to track and measure molecules in the body and environment.

Magic Numbers

There are certain numbers of protons and neutrons, called *magic numbers*, in which nuclei are stable. Scientists found common magic numbers of 2, 8, 20, 28, 50, 82, and 126 that match up with the same number of protons and neutrons. In other words, the simplest magic number "2" is a helium atom with two protons and two neutrons.

> **Magic numbers** are the number of protons and neutrons that make the most stable atomic nuclei.

Magic numbers demonstrate how some proton and neutron combinations are stable (i.e., nonradioactive), while others are unstable (i.e., radioactive).

Radioactive elements decay (lose energy) over time. During decay, an unstable nucleus tries to stabilize itself by emitting particles. This goes on in a stepwise way until a final, stable proton and neutron configuration is accomplished. These configurations are the magic numbers. For example, radioactive uranium (U-238) decays eventually to lead (Pb-82).

Magic number nuclei also have a higher than average binding energy per nuclear particle. This makes it possible for transuranium elements to be formed with extremely large nuclei. They don't fall victim to the super fast radioactive decay commonly experienced by isotopes with high atomic numbers. Large isotopes with magic numbers of nucleons are said to exist in an *island of stability*.

What Is Half-Life?

All radioactive isotopes decay. The amount of time it takes for one-half of a sample to decay is called the isotope's *half-life* (i.e., $t_{1/2}$). Different radioisotopes have different half-lives. These time periods don't depend on pressure, temperature, or bonding properties.

> **Half-life** is the time needed for half of the atoms of a radioactive isotope to undergo radioactive decay into a different form.

Depending on the half-life, there are varying amounts of radioactivity present in a sample. The following shows the relationship of half-life to percent radioactivity remaining in a sample:

No half-lives → 100% (i.e., no decay in a radioactive sample)
One half-life → 50%
Two half-lives → 25%
Three half-lives → 12.5%

This continues until the amount of radioactivity left in a sample is undetectable. In order to calculate the amount of a radioisotope left, the following equation is used:

$$N_t = N_o \times (0.5)^{\text{ number of half-lives}}$$

where N_t is radioisotope sample present, N_o is original radioisotope quantity, and number of half-lives is time divided by half-life.

EXAMPLE 15-6

Strontium-90 (^{90}Sr) has a half-life of around 28 years. So initially (t = 0 years),

100% ^{90}Sr present

28 years later, only half of the original amount of ^{90}Sr is left

½ × 100% = 50% ^{90}Sr

After 28 more years, only half of that ^{90}Sr is left

½ × 50% = 25% ^{90}Sr

After 28 more years, only half of that ^{90}Sr is left

½ × 25% = 12.5% ^{90}Sr

This radioactive decay pattern repeats until the ^{90}Sr level is undetectable from the earth's natural background radiation levels.

Longer radioactive decay periods are hard to imagine. For example, the half-life of $^{238}_{92}$U is 4.5×10^9 years. This is about the same as the age of the Earth. It is amazing to think that uranium found today will be around for another 4 billion years. Table 15-2 lists different radioactive decay rates for different elements.

TABLE 15-2 Radioactive elements decay at different rates

Element	Decay Rate (half-life)
Rhodium–106	30 seconds
Tellurium–134	42 minutes
Rhodium–103	57 minutes
Lanthanum–140	40 hours
Radon–222	4 days
Xenon–133	5 days
Iodine–131	8 days
Barium–140	13 days
Cerium–141	32 days
Niobium–95	35 days
Ruthenium–103	40 days
Strontium–89	54 days
Zirconium–95	65 days
Ruthenium–106	1 year
Cerium–144	1.3 years
Promethium–147	2.3 years
Krypton–85	10 years
Hydrogen–3	12 years
Curium–224	17.4 years
Strontium–90	28 years
Cesium–137	30 years
Plutonium–238	87 years
Americium–241	433 years
Radium–226	1622 years
Plutonium–240	6500 years
Americium–243	7300 years
Plutonium–239	24,400 years
Technecium–99	2×10^6 years
Iodine–129	1.7×10^7 years
Uranium–235	7.1×10^8 years
Uranium–238	4.5×10^9 years

Carbon Dating

Nonradioactive or common carbon is ^{12}C. A carbon isotope, C-14, created when atmospheric neutrons from cosmic radiation react with nitrogen atoms, has a half-life of 5730 years and is used to calculate the age of archeological objects.

Isotopes are used to date ancient soils, plants, animals, and the tools of early people. Because the radioactive decay rate of carbon-14 is constant, decay rate calculations can be used to find the number of years that have passed compared to the carbon-14 half-life. The carbon-14 decay is written as:

$$^{14}_{6}C \rightarrow {}^{14}_{7}N + 0/-1e \text{ (half-life is 5730 years)}$$

EXAMPLE 15-7

This example shows how radiocarbon (C-14) dating of an organic object (e.g., cotton fibers from an ancient burial site) is calculated. Radioactive decay takes place as the equation below:

$$\log_{10} X_0/X = kt/2.30$$

where X_0 is initial amount of radioactive material, X is the amount of material remaining after time t, and k is the first-order rate constant [i.e., isotope (C-14) undergoing decay]. This rate constant is related to half-life. Remember, the half-life for carbon-14 is 5730 years. The following equation is used to find the rate constant for solving this problem:

$$k = 0.693/t_{1/2}$$
$$k = 0.693/5730 \text{ years} = 1.21 \times 10^{-4}/\text{year}$$
$$\log X_0/X = [(1.21 \times 10^{-4}/\text{year}] \times t]/2.30$$
$$X = 0.795 \, X_0, \text{ so } \log X_0/X = \log 1.000/0.795 = \log 1.26 = 0.100$$
$$\text{then, } 0.100 = [(1.21 \times 10^{-4}/\text{year}) \times t]/2.30$$
$$t = 1900 \text{ years}$$

Carbon dating of organic materials depends a lot on the condition of the original sample and is generally considered accurate to between 30,000 and 50,000 years. Other assay methods using uranium, lead, potassium, and argon can measure much longer time periods, since they are not limited to once-living organic samples. The oldest rocks tested in the earth's crust have been dated to around 4 billion years. Radioactive dating also made it possible for scientists to date meteorites to around 4.4×10^9 years.

Radiation Exposure

As mentioned earlier, exposure to ionizing radiation can damage living tissue. With enough energy, ionizing radiation strips electrons away from atoms or breaks chemical bonds to create ions. An organism's defense mechanisms try to repair the damage, but sometimes it is too severe, widespread, or not repairable. So, it is important to know radiation exposure levels and damage. Table 15-3 gives common exposure doses from various sources.

Several factors influence possible health problems resulting from radiation exposure, but the main factors are:

1. Dose size (i.e., amount of radiation experienced by the body)

2. Susceptibility of a specific tissue to radiation

3. Organs affected

When measuring radiation exposure, different terms are used depending on what aspect of radiation is being studied. Radiation exposure affects some organs and tissues more than others.

TABLE 15-3 Dose size is affected by different variables		
Source	**Dose (rem)**	**Dose (sievert)**
Average person's whole-body exposure from the environment (e.g., soil, cosmic rays) in U.S.*	360 mrem per year	3.60 mSv
Average dose of nuclear power workers in the U.S.*	300 mrem per year (in addition to background dose)	3.00 mSv
One mammogram†	70 mrem	0.7 mSv0
Cosmic rays exposure during flight from Athens to New York†	63 mrem per 100 block hours	0.63 mSv
One chest x-ray†	10 mrem	0.1 mSv
One dental x-ray†	4-15 mrem	0.04-0.15 mSv
*EPA; Idaho State University (see Internet references). †Centers for Disease Control.		

Radiation Dosage

When exposed to radioactive elements, a person absorbs radiation. The conventional unit *rad* or the SI unit *gray* (Gy) is used to measure *radiation dosage* ; 1 Gy is equal to 100 rad. It is important for physicians treating a patient exposed to radiation to know dosage, since absorbed energy causes tissue damage. The more energy absorbed by cells, the higher the biological damage.

Measuring Biological Risk

Radiation exposure has biological risks, such as the development of cancerous cells. The conventional unit *rem* or the SI unit *sievert* (Sv) is used for estimating biological risk associated with exposure to radioactive compounds; 1 Sv is equal to 100 rem. The different forms of ionizing radiation have different biological exposure risks. Table 15-4 shows related radiation dose units.

Scientists developed a number, the quality factor (Q), to estimate biological risks from radiation exposure. When someone is exposed to radiation, physicians multiply the dose (i.e., rad) by the quality factor for the radiation type to get the biological risk for the person in rems. In other words, biological risk is:

$$\text{rem} = \text{rad} \times Q$$

Table 15-5 shows different doses and associated damage from radiation exposure.

TABLE 15-4 Radiation is measured in dose units (e.g., rem)			
Quantity	**SI Unit**	**Non–SI Unit**	**Conversion Factor**
Radioactivity	Becquerel (Bq)	Curie (Ci)	1 Ci = 3.7 × 10^{10} Bq = 37 Gigabecquerels (GBq) 1 Bq = 27 picocurie (pCi)
Absorbed dose	Gray (Gy)	Rad	1 rad = 0.01 Gy
Equivalent dose	Sievert (Sv)	Rem	1 rem = 0.01 Sv 1 rem = 10 mSv

TABLE 15-5 Radiation damage is related to dose levels

Dose (rem)	Consequences
5-20	Potential long-term effects and/or chromosomal damage
20-100	Brief drop in total white blood cell count
100-200	Mild radiation sickness within a few hours (e.g., vomiting, diarrhea, fatigue, hair loss at higher exposure)
200-300	Radiation sickness effects as above in 100-200 rem plus hemorrhage; lethal dose to 10-35% of the population after 30 days (LD 10-35/30)
300-400	Severe radiation sickness; bone marrow and intestine devastation; LD 50-70/30
400-1000	Acute illness, premature death; LD 60-95/30
1000-5000	Mortal illness, death in 100% of population in a few days; LD 100/10

Nuclear Medicine

Nuclear medicine pinpoints disease on the basis of metabolic changes, rather than structural changes usually detected by x-ray imaging in hard tissue (e.g., bones). After a nuclear medicine technologist gives a radiopharmaceutical to a patient by mouth, injection, inhalation, or other means, tissue and organ function are monitored.

Radiopharmaceuticals

Nuclear medicine is the branch of radiology that uses minute amounts of *radionuclides* (i.e., unstable atoms that emit radiation) to diagnose and treat disease. These radionuclides form *radiopharmaceuticals*, which make it possible to image an organ's structure and inner workings.

Radiopharmaceuticals eventually absorb into tissues and emit imaging signals. Radiation detectors take a picture of where the radiopharmaceuticals have been absorbed in the body and where they have not. The diagnostic images are sent to a computer screen and regions of higher-than-normal or lower-than-normal radioactive concentrations are flagged for analysis by a physician (i.e., radiologist) to study.

In the past 20 years, radiopharmaceuticals have been used in cancer detection and treatment. Radiology is often used to help diagnose and treat problems very early in a disease, (e.g., thyroid cancer). Because rapidly dividing cells like those in cancer are more susceptible to radiation than slow-growing normal cells, treatment using medical isotopes works well.

The radionuclide, cobalt-60, is useful in cancer therapy when implanted near a cancer.

Radiation therapy or radiotherapy can also be used by directing a narrow beam to an inoperable brain cancer. Other times, it can be used before surgery to shrink a cancerous lung or breast tumor.

> A **radioactive tracer** is a very small amount of radioactive isotope added to a chemical, biological, or physical system to study it.

Barium and thallium are radionuclides useful in understanding disease. Radioactive barium (Ba-37) is used to diagnose unusual abdominal pain, gastroesophageal reflux, and gastric or duodenal ulcers or cancer.

Thallium (Tl-201) is a radioactive tracer used to detect heart disease. This isotope binds tightly to well-oxygenated heart muscle. When a patient with heart trouble is tested, a scintillation counter detects the levels of radioactive thallium bonded to oxygen. Regions in the heart not receiving oxygen (i.e., with little to no thallium binding) are seen as dark areas.

Radioimmunology

You learned about biomarkers and how fluorescence can be used to identify disease conditions in Chap. 13. *Radioimmunology* measures the level of biological factors (e.g., proteins, enzymes) known to be affected by different diseases. Similar to fluorescent biomarkers, enzymes and proteins can also be tagged with radioactive elements. Some of the medical isotopes used in diagnosing disease include, barium (Ba-136), gallium (Ga-67), technetium (Tc-99), and iodine (I-123).

According to the World Nuclear Association, "Over 10,000 hospitals worldwide use radioisotopes in medicine, and about 90% of the procedures are for diagnosis. The most common radioisotope used in diagnosis is technetium-99, with some 30 million procedures per year, accounting for 80% of all nuclear medicine procedures worldwide."

Radioactive elements are an important area of advanced biochemistry. This area of research continues to grow as new imaging and detection methods are developed and more and more nuclear subparticles discovered. Nuclear medicine imaging is a combination of many different sciences, including analytical and physical chemistry, physics, mathematics, computer technology, and medicine.

Other Radioactive Element Use

When properly shielded, radioactive elements have many modern uses. Elements (e.g., thorium, curium, uranium) in the actinide series are used as power supplies for pacemakers, nuclear satellites, and submarines.

Americium-241 is used in home smoke detectors to test air conductivity. The isotope is located in an ionization chamber with a small, constant current between the electrodes. Any smoke entering the chamber absorbs the alpha particles, reduces the ionization, breaks the current, and sets off the alarm. Radioactive properties of americium-241 allow very small amounts of smoke to be detected.

Radioactive Waste

Radioactive waste resulting from energy-generating processes must eventually be stored safely. Since a radioactive element takes its own time to decay, the process can't be hurried.

Consequently, radioactive waste is a huge concern for many governments worldwide. The safe storage or disposal of radioactive wastes from nuclear power plants and atomic weapons is a top priority. Protecting populations from handling accidents or terrorist threats is critical. In fact, it drives advanced research and efforts to better understand reactivity and degradation of radioactive compounds and elements.

QUIZ

1. A nuclear medicine imaging technique, which produces a 3D image of functional processes in the body, is known as
   A. PEG.
   B. PET.
   C. PEA.
   D. PER.

2. Chemically identical atoms of the same element with different numbers of neutrons and mass numbers are called
   A. isomers.
   B. enantiomers.
   C. isotopes.
   D. chiral forms.

3. A very small amount of radioactive isotope added to a chemical, biological, or physical system to study it is called a
   A. calorimeter assay.
   B. radioactive tracer.
   C. dye marker.
   D. hydrogenated oil.

4. A beta (β) particle is an
   A. electron from the nucleus.
   B. electron that changes into antimatter.
   C. opaque form of carbon.
   D. electron from the electron cloud.

5. Which of the following is not used as a diagnostic tool in nuclear medicine?
   A. Barium
   B. Iodine
   C. Technetium
   D. Plutonium

6. Water formed from deuterium and oxygen is known as
   A. radioactive waste.
   B. heavy water.
   C. agua loca.
   D. protiated water.

7. **What is the half-life of ^{14}C?**

   A. 5730 years
   B. 7720 years
   C. 9520 years
   D. 10,400 years

8. **SI units for absorbed dose is written as**

   A. lumen units.
   B. ohm units.
   C. gray units.
   D. amps.

9. **Radioactive isotope decay**

   A. gives off emitted energy.
   B. can occur in seconds or thousands of years.
   C. can occur through alpha, beta, particle emissions.
   D. All of the above

10. **High-energy electromagnetic waves like light, but with a shorter, more penetrating wavelength are called**

    A. alpha particles.
    B. beta particles.
    C. gamma rays.
    D. sine waves.

Final Exam

1. **Researchers use rounding to**
   A. drop digits greater than 10.
   B. reduce precision, but increase accuracy.
   C. increase all numbers to the most certain number.
   D. drop nonsignificant digits in a calculation.

2. **Which natural atmospheric layer blocks the Earth from cancer-causing and damaging ultraviolet radiation from the sun?**
   A. Ozone
   B. Ionosphere
   C. Troposphere
   D. Exosphere

3. **Which of the following gases is not generally considered a major factor in global warming?**
   A. nitrous oxide
   B. methane
   C. carbon dioxide
   D. krypton

4. **When a proton grabs an electron from a previously filled ($\uparrow\downarrow$) energy level, creating an empty slot ($\uparrow\_$) in the (1s) energy level, it is known as**
   A. gamma emission.
   B. electron capture.
   C. alpha emission.
   D. electron pairing.

5. If 12 out of 80 chemistry students receive A's in a 6-month period, what percent receive A's in that period?

   A. 15%
   B. 22%
   C. 27%
   D. 33%

6. Atomic number (Z) equals the

   A. volume of the atomic sample.
   B. atomic weight of an element.
   C. sum of the reactants in a rate equation.
   D. number of protons in the atom's nucleus.

7. Observation, hypothesis, prediction, and experimentation are all parts of

   A. Le Châtelier's principle.
   B. scientific method.
   C. ideal gas law.
   D. Hund's rule.

8. Pico is the Greek prefix for the number

   A. 10^2.
   B. 10^6.
   C. 10^{-9}.
   D. 10^{-12}.

9. Which subatomic particles, orbiting nucleus like satellites, are attracted by the forces of electromagnetism?

   A. Electrons
   B. Photons
   C. Neutrons
   D. Light waves

10. The third period contains how many elements?

    A. 6
    B. 8
    C. 16
    D. 34

11. The standardized system used to name compounds is called its

    A. shorthand.
    B. reaction rate.
    C. common language.
    D. nomenclature.

12. **Isopropyl alcohol is an example of a**

    A. isotonic solution.
    B. miscible solution.
    C. ester.
    D. aromatic hydrocarbon.

13. **Stickstoff is the German name for which element?**

    A. Gold
    B. Sodium
    C. Nitrogen
    D. Selenium

14. **What kind of formula indicates the type and number of atoms that make up a compound?**

    A. Common
    B. Nuclear
    C. Reaction
    D. Molecular

15. **The study of organisms and systems at the molecular level (e.g., metabolism) is called**

    A. nuclear chemistry.
    B. physical chemistry.
    C. biochemistry.
    D. nanotechnology.

16. **Which early scientist used the atomic weight of elements to arrange 28 elements into six families with similar chemical and physical characteristics?**

    A. Alain Meunier
    B. Ernest Rutherford
    C. Patrick Drack
    D. Lothar Meyer

17. **The modern Periodic Table contains roughly how many elements?**

    A. 33
    B. 92
    C. 102
    D. 118

18. **Gold, silver, and copper are all what kind of elements?**

    A. Transition metals
    B. Nonconducting
    C. Nonmetals
    D. Radioactive

19. What is the molecular formula of galactose?

   A. $C_2H_2O_6$
   B. $C_6H_{10}O_2$
   C. $C_6H_{12}O_6$
   D. $C_{12}H_6O_2$

20. All halogens are

   A. phopholipids.
   B. nonmetals.
   C. chiral molecules.
   D. liquid at room temperature.

21. What is made from the combination of two or more metals or a metal and nonmetal?

   A. Biodegradable containers
   B. Acid rain
   C. Complex carbohydrates
   D. Alloys

22. When a copper penny floats on the surface of mercury, it is because of differences in

   A. surface temperature.
   B. color.
   C. density.
   D. interactions.

23. Amides are formed from carboxylic acids where the –OH group is replaced with a

   A. NH_2.
   B. CH_3.
   C. Cl_2.
   D. SO_4.

24. When a solution of a sample can't dissolve any more substance (i.e., no more reactions can occur), it is at its

   A. melting point.
   B. homologous series.
   C. vaporization point.
   D. saturation point.

25. Different structural forms of the same element are called

   A. purines.
   B. allotropes.
   C. radioactive decay.
   D. inner transition metals.

26. **The Periodic Table has a zigzag line that divides**
    A. gases.
    B. radioactive elements.
    C. metals and nonmetals.
    D. salts.

27. **The charge at the anode is**
    A. positive.
    B. zero.
    C. negative.
    D. changing too quickly to measure.

28. **Substances or solutions with free ions that are electrically conductive are called**
    A. signets.
    B. insulators.
    C. ⁻halogens.
    D. ⁻electrolytes.

29. **A mixture of salt and pepper is an example of a**
    A. one phase.
    B. heterogenous mixture.
    C. homologous structure.
    D. homogenous mixture.

30. **A semiconductor nanocrystal a few nanometers to a few hundred nanometers in overall size is called a**
    A. diamond.
    B. quantum dot.
    C. argon.
    D. fluorine.

31. **What is the atomic number of Nobelium?**
    A. 52
    B. 87
    C. 91
    D. 102

32. **Which disaccharide is formed from one molecule each of galactose and glucose?**
    A. Fructose
    B. Lactose
    C. Sucrose
    D. Maltose

33. Orbitals of the s-type come in
   A. singles.
   B. pairs.
   C. triplets.
   D. there are no s orbitals.

34. Which characteristic of a material equals the ratio of its density to the density of water at 4°C?
   A. Temperature
   B. Gravimetric pressure
   C. Specific gravity
   D. Enthalpy

35. The reaction, acid + base → salt + water, is known as a
   A. catalytic reaction.
   B. neutralization reaction.
   C. homologous reaction.
   D. synthesis reaction.

36. Molar heat of vaporization is the amount of heat it takes to vaporize 1 mol of liquid at a constant temperature and
   A. volume.
   B. concentration.
   C. pressure.
   D. humidity.

37. If it is −2°C in Glasgow, Scotland, in winter, what is that in Fahrenheit in Seattle, Washington?
   A. 9
   B. 15
   C. 28
   D. 32

38. The maximum number of electrons filling a specific energy level is found using which of the following formulas?
   A. $2n^2$
   B. $2(n + 1)$
   C. n^2
   D. $2n + n$

39. What is the capability of a liquid to flow or not flow freely at room temperature called?
   A. Viscosity
   B. Melting point
   C. Surface tension
   D. Triple point

40. **Chemical reactions in a solution at the interface between an ionic conductor and an electrical conductor describe which chemistry discipline?**

    A. Green chemistry
    B. Physical chemistry
    C. Electrochemistry
    D. Theoretical chemistry

41. **Sodium chloride is commonly called**

    A. lye.
    B. chitin.
    C. Epsom salt.
    D. table salt.

42. **Which of the following reactions shows the acid-base reaction of hydrobromic acid with sodium hydroxide?**

    A. $HBr + NaOH \rightarrow NaBr + OH$
    B. $HBr + NaOH \rightarrow Na + HBr + OH$
    C. $HBr + NaOH \rightarrow NaBr + H_2O$
    D. $HBr + NaOH \rightarrow HBr + NaO_2$

43. **If a solution's pH is less than 7.0, it is said to be**

    A. acidic.
    B. neutral.
    C. basic.
    D. without protons (H^+ ions).

44. **Nuclei with 2, 8, 20, 28, 50, 82, or 126 protons or neutrons have been found to be particularly stable and are called**

    A. radioactive.
    B. chiral.
    C. perfect numbers.
    D. magic numbers.

45. **The compound $CH_3COO(CH_2)_4CH_3$ (banana smell) is what type of organic compound?**

    A. Amine
    B. Alcohol
    C. Ester
    D. Lipid

46. **Light-emitting diodes (LED) used for medical applications (e.g., cancer treatment) came from**

    A. NOAA.
    B. NASA.
    C. EPA.
    D. USDA.

47. **How many protons, electrons, and neutrons are in $^{93}_{41}$Nb?**
    A. 93 protons, 41 electrons, and 0 neutrons
    B. 41 protons, 93 electrons, and 134 neutrons
    C. 41 protons, 41 electrons, and 93 neutrons
    D. 41 protons, 41 electrons, and 52 neutrons

48. **Along with oxygen, what isotope forms heavy water?**
    A. Deuterium
    B. Nitrogen
    C. Tritium
    D. Neon

49. **What is the element symbol for an atom with 22 protons and 26 neutrons?**
    A. Mg
    B. Ca
    C. Ti
    D. Au

50. **Oxidation is defined as**
    A. electron (+) gain.
    B. only occurring at pH = 8.
    C. general neutralization reaction.
    D. electron (−) loss.

51. **When gases expand and mix with other gases to fill available space, it is called**
    A. diffusion.
    B. expansion.
    C. carbonation.
    D. diffraction.

52. **What level of reaction is shown in the following rate equation, A → B (i.e., reaction rate = k [A])?**
    A. No reaction occurs
    B. First-order reaction
    C. Second-order reaction
    D. Third-order reaction

53. **If a person received a radiation dose of 300–400 rem, which of the following outcomes would they likely experience? (See Table 15-5)**
    A. Mild radiation sickness (fatigue, vomiting)
    B. Severe radiation sickness (bone marrow/intestinal devastation)
    C. Acute illness (premature death)
    D. Mortal illness (100% death in population)

54. **The equation used to find kinetic energy is**

    A. $P_iV_i = P_fV_f$.
    B. $V \propto 1/P$.
    C. $KE = \frac{1}{2}mv^2$.
    D. $r_1/r_2 = \sqrt{M_2/M_1}$.

55. **A manometer measures**

    A. color gradients.
    B. volume inside a sphere.
    C. temperature.
    D. the pressure inside a container.

56. **Which of the following is an alloy?**

    A. Silver.
    B. Brass.
    C. Copper.
    D. Nickel.

57. **When metals are malleable, they**

    A. are used in cleaning fluids.
    B. can be hammered into flat sheets.
    C. can be pulled into thin wires.
    D. are used in structural supports.

58. **A solution is a homogeneous mixture of a solute (i.e., element or compound being dissolved) in a**

    A. gas.
    B. amino acid.
    C. solvent.
    D. salt mixture.

59. **If a solutions temperature is 21°F, what is the temperature in kelvin?**

    A. 217
    B. 235
    C. 253
    D. 267

60. **Any material able to be broken down into simpler components by microorganisms is**

    A. immiscible.
    B. a catalyst.
    C. biodegradable.
    D. solid at room temperature.

61. Today's CO_2 levels are the highest they've been over the past 650,000 years, nearly

    A. 200 ppm.
    B. 300 ppm.
    C. 400 ppm.
    D. 500 ppm.

62. These following elements are all metalloids, except

    A. antimony.
    B. germanium.
    C. arsenic.
    D. strontium.

63. The pH of cow's milk is roughly

    A. 5.2.
    B. 6.5.
    C. 8.1.
    D. 9.0.

64. The amount of potential work available from an isothermal, isobaric thermodynamic system is called

    A. enantiomeric process.
    B. matter transfer.
    C. Gibbs free energy.
    D. Arrhenius rate equation.

65. Cordless tools and memory foam are examples of

    A. ThinkGeek.com™ gadgets.
    B. medical breakthroughs.
    C. winning science fair projects.
    D. NASA spin-offs.

66. Avogadro's number is equal to

    A. 2.022×10^{23}.
    B. 6.022×10^{23}.
    C. 7.025×10^{13}.
    D. 8.022×10^{23}.

67. A lava lamp provides an example of

    A. biochemistry at its finest.
    B. miscible solutions.
    C. pressure effects on solutions.
    D. immiscible solutions.

68. **Which of the following is a correct measurement?**

    A. Ohms per atmosphere
    B. Kilometers per bushel
    C. Degrees Celsius
    D. Joules per liter

69. **Carbon's atomic number is**

    A. 6.
    B. 8.
    C. 12.
    D. the same as its atomic weight.

70. **The SI standard unit for mass is written in**

    A. atmospheres.
    B. drams.
    C. kegs.
    D. kilograms.

71. **In van der Waals' equation, $(P + n^2a/V^2)(V - nb) = nRT$, a stands for**

    A. temperature.
    B. pressure.
    C. a constant which accounts for attractive forces between molecules.
    D. a constant which accounts for the volume of each molecule.

72. **The universal solvent is**

    A. alcohol.
    B. water.
    C. mercury.
    D. molasses.

73. **The spins of the two electrons in each filled orbital are**

    A. slower than for unfilled orbitals.
    B. the same.
    C. opposite in direction.
    D. of the same charge.

74. **If a solution's pH is greater than 7.0, it is said to be**

    A. acidic.
    B. basic.
    C. neutral.
    D. without charge.

75. **In Charles' law, when pressure is held constant, a volume of gas is directly proportional to the**
    A. atmosphere.
    B. acceleration.
    C. pH.
    D. kelvin temperature.

76. **Light scattered by small particles in suspension (e.g., colloid) is called the**
    A. Tyndall effect.
    B. Tsai constant.
    C. de Broglie wavelength.
    D. McSweeney diffraction.

77. **The time it takes for half the atoms of a radioactive isotope to decay into a different radioactive form is the**
    A. dosage rate.
    B. half-life.
    C. tracer rate.
    D. number of beta particles.

78. **The SI standard unit to measure a pure substance is the**
    A. joule.
    B. nanometer.
    C. mole.
    D. gram.

79. **In a simple Periodic Table, besides the element symbol, what other piece of information is always given?**
    A. Melting point
    B. Electronegativity number
    C. Heat of fusion
    D. Atomic number

80. **The sum of the atomic weights of atoms in a molecule of a substance is known as**
    A. Planck's constant.
    B. Dalton's law of partial pressures.
    C. atomic weight.
    D. enthalpy.

81. **Which chemistry discipline focuses on the rate at which chemical reactions take place?**
    A. Organic chemistry
    B. Combinatorial chemistry
    C. Photochemistry
    D. Kinetics

82. **The word radioactivity was first used by**

    A. Lothar Meyer.
    B. Marie Curie.
    C. Vernon Hill.
    D. Joseph Gay-Lussac.

83. **$PV = nRT$ is the formula for**

    A. Hund's rule.
    B. Le Châtelier's principle.
    C. ideal gas law.
    D. Avogadro's law.

84. **Parts per billion is commonly used to measure**

    A. pollutants.
    B. turbidity.
    C. radiation.
    D. sand particles.

85. **The pH of a common stomach antacid is roughly**

    A. 6.5.
    B. 7.2.
    C. 10.
    D. 12.7.

86. **Different types of atoms with the same atomic number, but different atomic mass are called**

    A. conductors.
    B. enzymes.
    C. elastomers.
    D. isotopes.

87. **The electronegativity value of cobalt (see Table 8-1) is**

    A. 0.7.
    B. 1.3.
    C. 1.8.
    D. 2.3.

88. **Particles dispersed through a gas are called a gas**

    A. constant value.
    B. aerosol.
    C. emulsion.
    D. irritant.

89. The attachment between atoms within a molecule is called a
    A. romance.
    B. symbiotic relationship.
    C. chemical bond.
    D. chance happening.

90. The idea that every orbital in a subshell is singly occupied with one electron before any one orbital is doubly occupied is known as
    A. Charles's law.
    B. Caputo's rule.
    C. Dalton's law.
    D. Hund's rule.

91. If *M* equals the number of moles of solute (*n*) per volume in liters (*V*) solution, what is *M*?
    A. Measurement
    B. Molarity
    C. Mass
    D. Moment

92. The biological molecules in living cells, organs, systems, and the environment are divided into four types: proteins, carbohydrates, lipids, and
    A. transuranic elements.
    B. tar.
    C. nucleic acids.
    D. chitin.

93. Which of the following is an aromatic compound?
    A. Vanillin
    B. Propane
    C. Ethane
    D. Methylene

94. Radioactive barium is used as a radioactive tracer to diagnose
    A. flatulism.
    B. halitosis.
    C. ulcers.
    D. cavities.

95. These compounds are secreted by organisms (e.g., bees) to attract mates, signal danger, or pass along directions
    A. fatty acids.
    B. β-carotene.
    C. cellulose.
    D. pheromones.

96. These molecules keep particles mixed instead of letting them settle back into separate layers.

   A. Carbohydrates
   B. Amino acids
   C. Surfactants
   D. Nanocrystals

97. $P_{total} = P_1 + P_2 + P_3 + P_4 + P_5 + \cdots$ is the equation for

   A. Boyle's law.
   B. Combined gas law.
   C. Charles's law.
   D. Dalton's law of partial pressures.

98. In molecular geometry, the nearness of electrons pairs to each other is best described by

   A. conversion factors.
   B. valence shell electron-pair repulsion (VSEPR).
   C. a phase diagram.
   D. nuclear isotope formation.

99. This instrument is designed to measure the amount of hydrogen ions in a solution.

   A. pH meter
   B. Scintillation counter
   C. Turbidity meter
   D. Geiger counter

100. In the reaction ($RCOOH + NaOH \rightarrow$) which of the following give the correct reaction products?

   A. $RCO_2 + NaOH$
   B. $R + COOH + Na$
   C. $RCOONa + H_2O$
   D. $NaOH + H_2O$

101. Phenol is a common name for

   A. amino acid.
   B. hydroxybenzene.
   C. bleach.
   D. inorganic salt.

102. A nanometer is equal to

   A. 1×10^{-4} meters.
   B. 1×10^{-7} meters.
   C. 1×10^{-9} meters.
   D. 1×10^{-10} meters.

103. **Platinum is located in which period of the Periodic Table?**
    A. Third
    B. Fourth
    C. Fifth
    D. Sixth

104. **If the electron capacity of an energy level is $2n^2 = 18$, what is n?**
    A. $n = 1$
    B. $n = 2$
    C. $n = 3$
    D. $n = 4$

105. **Gay-Lussac's law, when volume is held constant, the pressure of a gas is directly proportional to its**
    A. kelvin temperature.
    B. color.
    C. mass.
    D. pH.

106. **The most common type of covalent bonds form between**
    A. two nonmetals.
    B. two metals.
    C. two solids.
    D. a metal and a nonmetal.

107. **If the temperature of a sample is 24 kelvin, what is its temperature in Celsius?**
    A. $-100.50\ °C$
    B. $-198.34\ °C$
    C. $-249.15\ °C$
    D. $-275.21\ °C$

108. **Which of the following is not a purine or pyrimidine base?**
    A. Thymine
    B. Uracil
    C. Guanine
    D. Citrine

109. **A double-bonded hydrocarbon is known as an**
    A. alkane.
    B. alkene.
    C. alkyne.
    D. aldehyde.

110. In the equation for calorimetry (i.e., $q = m \times c \times \Delta T(T_{final} - T_{initial})$, what does the variable "c" represent?

    A. Energy
    B. Temperature
    C. Specific heat capacity
    D. Mass

111. The horizontal rows (numbers 1–7) on the left-hand side of the Periodic Table are called

    A. groups.
    B. isomers.
    C. enantiomers.
    D. periods.

112. Which of the following numbers represents 1.0723 rounded to the nearest hundredth?

    A. 1.072
    B. 1.07
    C. 1.8
    D. 2.0

113. Small molecules containing an amino group ($-NH_2$) and a carboxyl ($-COOH$) group are called

    A. unsaturated fats.
    B. phenols.
    C. saturated fats.
    D. proteins.

114. What is the approximate pH of battery acid?

    A. 0.8
    B. 2.2
    C. 4.9
    D. 7.0

115. Which of the following defines a chemical isomer?

    A. different formulas and lengths
    B. same molecular formulas, but different 3D shape
    C. same number of carbon to carbon bonds
    D. contain two oxygen atoms

116. Carbon-14 is used to

    A. make cookies.
    B. power solar panels.
    C. find the age of once-living materials.
    D. test for stomach cancer.

117. **Which of the following completes the electron filling pattern: 1s, 2s, 2p, 3s, 3p, 4s, 3d, ...?**

   A. 4d, 5p, 6s, 4f, 5d, 6p, 7s, 5f
   B. 4p, 5s, 6s, 4f, 5d, 6p, 7s, 5f
   C. 4d, 5s, 4p, 5p, 6s, 4f, 5d, 6p, 7s, 5f
   D. 4p, 5s, 4d, 5p, 6s, 4f, 5d, 6p, 7s, 5f

118. **What is the equilibrium of temperature and pressure of a substance's three phases on a phase diagram called?**

   A. Trimeric bonding
   B. Triple point
   C. Solubility quotient
   D. Melting point

119. **The ability of an atom to attract electrons in a chemical bond is called**

   A. isomerism.
   B. heat of reaction.
   C. electronegativity.
   D. Boyle's law.

120. **The IUPAC uses B > Si > C > P > N > H > S > I > Br > Cl > O > functional group when naming**

   A. ionic compounds.
   B. binary covalent compounds.
   C. electrolytes.
   D. carbon-carbon triple bonds.

121. **In the dilution $(M_{initial}) (V_{initial}) = (M_{final}) (V_{final})$ equation, if V represents volume then what does M represent?**

   A. Molarity
   B. Mass
   C. Momentum
   D. Metal

122. **The SI unit for measuring the rate of flow of electric charge (i.e., electrical current) is the**

   A. ohm.
   B. alternating current.
   C. ampere.
   D. kilogram/kilometer.

123. **Which formula represents dinitrogen trioxide?**

   A. NH_2
   B. HCO_3
   C. N_3HO_3
   D. N_2O_3

124. **Enantiomers or mirror image molecules are**

    A. always radioactive.
    B. superposable.
    C. nonsuperimposable.
    D. identical.

125. **In Boyle's law, when pressure increases (↑) then the volume of a gas**

    A. increases.
    B. decreases.
    C. doesn't change.
    D. zero.

126. **Which of the following contains a benzene ring?**

    A. Salt
    B. Oil of cinnamon
    C. Propane
    D. Water

127. **The compound NH_3 is commonly known as**

    A. vinegar.
    B. ammonia.
    C. bleach.
    D. lye.

128. **If a solution measured 5.7 on the pH meter, it would be**

    A. a radioactive isotope.
    B. a gas.
    C. a base.
    D. an acid.

129. **The density of ice is less than the density of water because**

    A. it is more transparent in color.
    B. it is more tightly packed.
    C. hydrogen bonding leaves empty spaces.
    D. oxygen is an atmospheric gas.

130. **Enthalpy is a state function and can be calculated using**

    A. Hess' law.
    B. Hund's rule.
    C. Drack's diagram.
    D. Dalton's law.

131. If the temperature of a sample is 99°F, what is the temperature in Centigrade?
    A. 37°C
    B. 40°C
    C. 53°C
    D. 76°C

132. Early alchemists wanted to turn common metals into
    A. platinum.
    B. gold.
    C. uranium.
    D. silver.

133. Johannes Brønsted and Thomas Lowry are best known for describing
    A. electron magnetism.
    B. metalloids.
    C. acids and bases.
    D. inert gases.

134. A triple-bonded hydrocarbon is known as an
    A. alkane.
    B. alkene.
    C. alkyne.
    D. aldehyde.

135. The least amount of energy needed for a chemical reaction to take place is called
    A. activation energy.
    B. First law of thermodynamics.
    C. effusion energy.
    D. Hund's rule.

136. DNA is the shortened name for which molecule?
    A. Diethylnitrosamine
    B. Dinitroglycerin
    C. Diamond acid
    D. Deoxyribonucleic acid

137. The oxidation state of an uncombined atom is
    A. >0.
    B. $^-1$.
    C. $^-2$.
    D. $^+1$.

138. **What gas is used to lift modern blimps (e.g., Goodyear® blimp)?**

    A. Neon.
    B. Hydrogen.
    C. Krypton.
    D. Helium.

139. **Which French chemist named oxygen and hydrogen, but was executed during the French Revolution because he was associated with tax laws?**

    A. Antoine Lavoisier
    B. Henry Louis Le Châtelier
    C. Antoine Henri Becquerel
    D. Jean-Luc Picard

140. **Global warming is having the biggest impact on**

    A. the price of propane.
    B. melting of polar ice.
    C. global magnetism.
    D. life span of butterflies.

141. **When two simple sugars combine, the product is called a**

    A. lipid.
    B. disaccharide.
    C. polysaccharide.
    D. monosaccharide.

142. **The delivery of drug molecules in the body where they are needed most is called**

    A. bioluminescence.
    B. biomarkers.
    C. biological coefficient.
    D. bioavailability.

143. **Iron is found in which group of the Periodic Table?**

    A. 2
    B. 4
    C. 8
    D. 14

144. **Triplets of nucleotide bases are called**

    A. plantains.
    B. codons.
    C. lipids.
    D. snippets.

145. Proteins that lower the activation energy of biochemical reactions are called
    A. enzymes.
    B. deactivators.
    C. fatty acids.
    D. diatomic molecules.

146. The amount of energy needed to raise the temperature of a set amount of sample 1°C is called
    A. a coulomb unit.
    B. heat capacity.
    C. density.
    D. excitation energy.

147. Chitin is a complex carbohydrate found in a
    A. unicorn's horn.
    B. human knee cap.
    C. daffodil plant.
    D. lobster shell.

148. A substance with a hydrophilic head and a hydrophobic tail is a
    A. saturated fat.
    B. surfactant.
    C. structural isomer.
    D. unsaturated fat.

149. Precision is related to
    A. how close a single measurement is to its true value.
    B. an element's atomic number.
    C. how carefully an amount is measured on an electronic balance.
    D. the closeness of two sets of measured values to each other.

150. What is the bond called that forms between two molecules where a carboxyl group from one molecule reacts with the amine group of another molecule, releasing water?
    A. Monatomic bond
    B. Radioactive bond
    C. Peptide bond
    D. Inorganic bond

Answers to Quizzes and Final Exam

Chapter 1	Chapter 3	Chapter 5	Chapter 7
1. B	1. C	1. D	1. C
2. C	2. C	2. B	2. B
3. D	3. B	3. A	3. D
4. C	4. D	4. C	4. A
5. A	5. A	5. D	5. C
6. D	6. C	6. A	6. A
7. C	7. A	7. B	7. D
8. D	8. C	8. C	8. B
9. A	9. A	9. C	9. B
10. B	10. D	10. D	10. C

Chapter 2	Chapter 4	Chapter 6	Chapter 8
1. B	1. B	1. C	1. D
2. C	2. C	2. D	2. C
3. D	3. B	3. B	3. A
4. D	4. A	4. B	4. A
5. C	5. D	5. B	5. B
6. C	6. C	6. C	6. C
7. A	7. A	7. A	7. D
8. B	8. C	8. A	8. B
9. C	9. A	9. D	9. A
10. D	10. D	10. D	10. B

Chapter 9
1. C
2. D
3. A
4. D
5. B
6. A
7. C
8. C
9. D
10. A

Chapter 10
1. A
2. C
3. B
4. C
5. B
6. C
7. D
8. B
9. A
10. C

Chapter 11
1. C
2. B
3. D
4. A
5. C
6. A
7. C
8. D
9. B
10. C

Chapter 12
1. C
2. D

3. B
4. C
5. D
6. A
7. B
8. C
9. B
10. D

Chapter 13
1. D
2. C
3. B
4. B
5. D
6. C
7. B
8. D
9. A
10. C

Chapter 14
1. C
2. D
3. B
4. A
5. B
6. D
7. C
8. C
9. D
10. B

Chapter 15
1. B
2. C
3. B
4. A
5. D

6. B
7. A
8. C
9. D
10. C

Final Exam
1. D
2. A
3. D
4. B
5. A
6. D
7. B
8. D
9. A
10. B
11. D
12. B
13. C
14. D
15. C
16. D
17. D
18. A
19. C
20. B
21. D
22. C
23. A
24. D
25. B
26. C
27. C
28. D
29. B
30. B
31. D
32. B

33. B
34. C
35. B
36. C
37. C
38. A
39. A
40. C
41. D
42. C
43. A
44. D
45. C
46. B
47. D
48. A
49. C
50. D
51. A
52. B
53. B
54. C
55. D
56. B
57. B
58. C
59. D
60. C
61. C
62. D
63. B
64. C
65. D
66. B
67. D
68. C
69. A
70. D

71. C	91. B	111. D	131. A
72. B	92. C	112. B	132. B
73. C	93. A	113. D	133. C
74. B	94. C	114. A	134. C
75. D	95. D	115. B	135. A
76. A	96. C	116. C	136. D
77. B	97. D	117. D	137. A
78. C	98. B	118. B	138. D
79. D	99. A	119. C	139. A
80. C	100. C	120. B	140. B
81. D	101. B	121. A	141. B
82. B	102. C	122. C	142. D
83. C	103. D	123. D	143. C
84. A	104. C	124. C	144. B
85. C	105. A	125. B	145. A
86. D	106. A	126. B	146. B
87. C	107. C	127. B	147. D
88. B	108. D	128. D	148. B
89. C	109. B	129. C	149. D
90. D	110. C	130. A	150. C

SI Base Units and Conversions

Length—SI base unit = meter (m)	
1 centimeter (cm)	0.39 inch (in)
1 inch (in)	2.54 centimeters (cm)
1 meter (m)	3.28 feet (ft); 1.0936 yards (yd)
1 foot (ft)	0.30 meter (m)
1 yard (yd)	0.91 meter (m)
1 kilometer (km)	0.62 statute mile; 3281 ft
1 fathom (fath)	6 feet; 1.83 meters (m)
1 angstrom (Å)	1×10^{-10} meter (m)
1 micrometer (µm)	1×10^{-6} m = 1 micron
1 nanometer (nm)	1×10^{-9} m (1 nm = 10 Å)

Area—SI base unit = square meter (m^2)	
1 square centimeter	0.16 square inch
1 square inch	6.45 square centimeters
1 square meter	10.76 square feet; 1.1960 square yards
1 square foot	0.09 square meter
1 square kilometer	0.39 square mile

Volume—SI base unit = cubic meter (m³)

1 cubic centimeter	0.06 cubic inch
1 cubic inch	16.39 cubic centimeters
1 cubic meter	35.31 cubic feet
1 cubic foot	0.028 cubic meter
1 cubic meter	1.31 cubic yards
1 liter	10^3 milliliters
1 gallon	3.79 liters

Mass—SI base unit = kilogram (kg)

1 gram (g)	0.06 ounce (oz); 10^3 milligrams (mg)
Mass density	1 kg/m³
1 ounce (oz)	28.3 grams (g)
1 kilogram (kg)	2.2 pounds (lb)
1 pound (lb)	454 grams (g)
1 metric ton (t)	10^3 kilograms (kg)

Pressure—SI base unit = pascal (Pa)

1 kilogram/cm²	0.96784 atmosphere; 14.2233 pounds/in²
1 kilogram/cm²	0.98 bar
1 bar	0.99 atmosphere; 10^5 pascals
1 atm	760.00 millimeter of mercury (mm Hg)

Temperature—SI base unit = kelvin (K)

Celsius to Fahrenheit	$°F = (9/5)°C + 32$
Fahrenheit to Celsius	$°C = (5/9)(°F - 32)$
Celsius to kelvins	$K = °C + 273.15$

Energy—SI base unit = joule (J)

1 joule (J)	1 kg m²/s² = 1 coulomb volt
1 calorie (cal)	4.18 joules (J)
1 food calorie (Cal)	1 kilocalorie (kcal) = 4184 joules (J)
1 British thermal unit (Btu)	252 calories (cal) = 1053 joules (J)

Electricity—SI base unit = ampere (A)	
Electric charge	SI unit for charge = coulomb (C), or ampere SI base unit
Electric current	SI unit for electrical current = ampere (A) = 1 coulomb/second
Electric resistance	SI unit of resistance = ohm (Ω)
Electric conductance	SI unit = siemens (S) (i.e., ratio of electric current and electric potential)
Electric potential	SI unit = voltage (V) = watt/ampere
Capacitance	SI unit = farad (F) (i.e., coulomb per voltage in terms of SI derived unit)

SI prefixes	
Prefix	**Value**
pico (p)	0.000000000001 *or* 10^{-12}
nano (n)	0.000000001 *or* 10^{-9}
micro (μ)	0.000001 *or* 10^{-6}
milli (m)	0.001 *or* 10^{-3}
centi (c)	0.01 *or* 10^{-2}
deci (d)	0.1 *or* 10^{-1}
deka (da)	10 *or* 10^{1}
hecto (h)	100 *or* 10^{2}
kilo (k)	1000 *or* 10^{3}
Mega (M)	$1,000,000$ *or* 10^{6}
Giga (G)	$1,000,000,000$ *or* 10^{9}
Tera (T)	$1,000,000,000,000$ *or* 10^{12}

Glossary

Accuracy Linked to how close a single measurement is to its true value.

Achiral Superimposable molecules.

Acid rain Includes all types of precipitation (rain, snow, sleet, hail, fog) that are acidic (pH lower than 5.6 average of rainwater) in nature.

Activation energy Lowest amount of energy needed for a chemical reaction to take place.

Adiabatic process In effect when a system of interest is thermally isolated so that no heat can enter or exit.

Alkanes Hydrocarbons with only single bonds.

Alkenes Hydrocarbons with at least one double bond.

Alkynes Hydrocarbons with at least one triple bond.

Amides Formed from carboxylic acids, but instead of the –COOH group, in an amide the –OH group is replaced by an –NH_2 group (e.g., –$CONH_2$).

Amines Primary amines (1°) have one hydrocarbon group bonded to an amino group (–NH_2); secondary amines (2°) have two hydrogen atoms replaced by hydrocarbon groups; and so on.

Amino acids Make up proteins; basic building blocks of enzymes, hormones, proteins, and body tissues; R–$CH(NH_2)COOH$.

Amphiprotic An ion or molecule which can either lose or add a proton (H^+) in a reaction.

Amphoteric An ion or molecule that can serve as either an acid or a base in a reaction, but has no protons.

Anabolism   A reaction where a group of molecules binds to form a larger molecule.

Anion   When an atom or molecule gains an electron, a negatively charged *anion* is formed.

Anode (−)   Electrode where oxidation takes place.

Aromatic hydrocarbons   Hydrocarbons with ring structures [e.g., *vanillin* $(C_8H_8O_3)$].

Arrhenius rate equation $[\ln (k) = E_a / R (1/T) + \ln (A)]$ Gives the rate constant of a chemical reaction to the exponential value of the temperature; ↑ the temperature, ↑ the rate constant.

Atomic number (Z)   Is equal to the number of protons in the nucleus of an atom.

Atomic symbol   A 1-, 2-, or 3-letter shorthand name for a specific element in the Periodic Table.

Atomic weight   Average atomic mass for a naturally occurring element in atomic mass units.

Aufbau principle   Also called the building principle; shows an atom's ground state.

Avogadro's law   States that equal volumes of any two gases at equal temperature and pressure contain an equal number of molecules.

Avogadro's number   A constant number of atoms, ions, or molecules in a sample is equal to the number of atoms in 12 grams of carbon-12 or 6.022×10^{23} (i.e., 1 mole).

Bioavailability   Delivery of drug molecules in the body where they are needed and will do the most good.

Biodegradable   A material which can be broken down into simpler components by microorganisms (e.g., bacteria).

Biological oxygen demand (BOD)   Chemical test for finding the amount of dissolved oxygen needed by biological marine organisms to break down organic material in a given water sample at a set temperature over a specific time.

Bioluminescence   Produces light via an enzymatic reaction (e.g., firefly's light).

Biomarker   Used to detect the presence of disease, an irregular characteristic, or a physiological condition; also called molecular markers or signature molecules.

Boyle's law   Temperature is held constant, a volume of gas is inversely proportional to the pressure; $V \propto 1/P$.

Brønsted acid   Proton donors.

Brønsted base   Any substance that can accept a hydrogen ion.

Buffers   Bind and release protons to lessen changes in pH in a solution.

Buffers   Are compounds used to react with hydrogen ions (H^+) and hydroxide ions (OH^-) which have neutralizing affects.

Calorie   1 calorie (1 cal) = 4.184 joules (4.184 J).

Calorimetry   The measurement of heat loss or gain during a chemical reaction, change of state (e.g., liquid to solid).

Carbohydrates   A large group of organic compounds containing carbon, oxygen, and hydrogen; made up of building blocks called simple sugars or monosaccharides; (e.g., fructose).

Carboxylic acids   Contain a –COOH group, which can be bonded to a hydrogen atom or an alkyl group.

Catabolism   A reaction in which a group of molecules are broken apart in smaller parts.

Catalyst   A small amount of a substance, added to reactants, which boosts the reaction rate without being consumed in the process; provide an extra reaction path with a lower activation energy.

Cathode (+)   Electrode where reduction takes place.

Cation   When an atom or molecule loses an electron, a *cation* is formed.

Charged species   A chemical component in which the overall electrons total is unequal to the overall protons total.

Charles's law   Pressure is held constant, a volume of gas is directly proportional to the kelvin temperature; $V \propto T$.

Chemical bond   Force that binds two or more atoms together.

Chemical oxygen demand (COD)   Chemical test used to check the amount of organic compounds in water.

Chirality   Property of non-superimposable molecules that exhibit "handedness" as mirror images; also called optical isomers.

Chromophore   A functional group capable of specific light absorption, creating color in a substance, and binding with other groups to form dyes; bonding allows it to absorb certain visible light wavelengths, while reflecting others.

Codons   Triplets of nucleotide bases.

Coenzyme   A small organic molecule that transports chemical groups between enzymes.

Colloid   A homogeneous solution made up of larger particles of one solution mixed and spread all through another solution.

Combined gas law   Made up of Boyle's, Charles's, and Gay-Lussac's laws to describe what happens when conditions for a gas changes (e.g., temperature, volume, and pressure); $P_i V_i / T_i = P_f V_f / T_f$.

Conjugate acid   Of a compound or ion forms when a proton is gained.

Conjugate acid-base pairs   Pairs of Brønsted-Lowry acids and bases such as H_2O, OH^- and NH_3, NH_4^+.

Conjugate base   Of a compound or ion forms when a proton is lost.

Contaminant   A substance in the environment at higher than normal background levels.

Contaminant receptor   Material or organism affected by a pollutant is called a receptor.

Contaminant sink   A contaminant sink is a material or species that stores and interacts with a pollutant.

Conversion factors   Use the relationship between two units or quantities expressed in fractional form.

Covalent bond   Electrons are shared equally between atoms.

Cracking   The breakdown of large hydrocarbon compounds into smaller compounds.

Dalton's law of partial pressures   When a gas is mixed with one or more different gases, the pressures of each gas add together to get the mixture's total pressure; $P_{total} = P_1 + P_2 + P_3 + \cdots$.

Deoxyribonucleic acid (DNA)   Made up of nucleic acids; a series of purine and pyrimidine bases [e.g., adenine (A), guanine (G), cytosine (C), thymine (T)].

Diatomic molecules   Molecules occurring naturally as two-atom molecules (e.g., oxygen, nitrogen, hydrogen, fluorine, chlorine, bromine, and iodine) occur in pairs at room temperature; grouped in IA and VIIA of the Periodic Table.

Diffusion   Occurs when gases expand and mix with other gases to fill available space.

Dimer   Two connected monosaccharides form this structure.

Dipole moment   Measurement of charge separation between each part of a molecule.

Dispersed phase   Substance in a colloidal mixture in the lesser amount.

Dispersing medium   Substance in a colloidal mixture in the greater amount.

Dissolution   Occurs when water molecules, acids, or other environmental compounds attract and remove oppositely charged ions from rock.

Dissolved oxygen   Amount of oxygen measured in a stream, river, lake, or other body of water.

Effusion   Leaking of a molecule (e.g., gas) from a closed container through a tiny hole.

Electrochemical cell   Provides electricity via chemical reactions.

Electrolysis   A chemical decomposition reaction created by passing electricity through an ion-containing solution.

Electrolytes   Any substance or solution with free ions making the substance electrically conductive.

Electromotive force (emf)   Force measured in volts created through the interaction between an electrical current and a magnetic field, one or more of which is changing; expresses electrical effect of a changing magnetic field.

Electron capture   Takes place in an energy level (1s) closest to the nucleus; a proton grabs an electron from a previously filled ($\uparrow\downarrow$) energy level, creating an empty slot ($\uparrow\_$) in the (1s) energy level.

Electronegativity   Ability of an atom or group to attract electrons.

Electrons   Small negatively charged subatomic particles that orbit around an atom's positively charged nucleus by forces of electromagnetism.

Electropositivity   Ability of an atom or group to give up electrons to other atoms or groups.

Electrostatic interactions   When ions' opposite charges attract one another.

Emulsifying agent   Is soluble in oil *and* water which allows it to mix the two; mustard allows oil and vinegar to mix in salad dressing.

Enantiomers   Stereoisomers that are mirror images of each other and non-superimposable.

Endothermic   ($\Delta H > 0$) = heat $\downarrow$ (absorbed), temperature $\downarrow$.

Enthalpy (H)   Amount of heat a sample contains; the amount of heat expelled during a reaction is ΔH; kJ mol^{-1} (kJ/mol) or kcal mol^{-1} (kcal/mol); to measure an enthalpy change at constant pressure, find the sum of the change between the products and reactants; $\Delta H = \Sigma H_{(products)} - \Sigma H_{(reactants)}$.

Entropy   Measure of unavailable energy within a closed system (e.g., the universe); also a measure of randomness or chaos within a closed system.

Enzymes   Proteins that lower the activation energy or catalyze roughly 4000 different biochemical reactions.

EPR effect   That is, enhanced permeability and retention; the process of super small injectable gold nanoshells traveling through a tumor's leaky vessels, to target and bind to diseased cells.

Equilibrium constant (K)   Is found by setting the reaction rates and chemical affinities for forward and backward reactions as equal.

Equivalence point   When titrating, the point where the solution is in equilibrium.

Esterification   Takes place when an organic acid reacts with an alcohol to create an ester (e.g., triphenyl phosphate) and water.

Esters   A class of organic compounds formed by combining acids and alcohols and losing a molecule of water in the process.

Exothermic   ($\Delta H < 0$) = heat $\uparrow$ (lost), temperature $\uparrow$.

Experiment   Controlled testing of the properties of a substance or system through carefully recorded measurements.

Factor-label method   Also known as *dimensional analysis*; changes one unit to another by using conversion factors.

Fats   Compounds (i.e., oils) made up of glycerol and three fatty acids bonded in a structure called a triglyceride.

Fatty acids   Made up of a hydrophobic "tail" (i.e., a long hydrocarbon chain) and a hydrophilic "head" of a carboxyl group; building blocks of fat molecules.

Fermentation   Anaerobic (no oxygen) conversion of organic compound (e.g., sugar) from grain into carbon dioxide and alcohol through reaction with yeast.

First law of electrolysis   The mass of a substance at an electrode is directionally propotional to the amount of electity passed through a solution.

First law of thermodynamics   Matter/energy is neither created nor destroyed, but remains constant in the universe (also called the law of conservation of matter).

Fluid dynamics   Tests the miscibility of fluids and the way they flow at different concentrations.

Fluorescence   Comes from light absorption by molecules called fluorophores.

Functional groups   Various groups added to hydrocarbons which give the molecule many different functions or properties.

Gay-Lussac's law   Volume is held constant, a pressure of a gas is directly proportional to the kelvin temperature in: $P \propto T$.

Gibbs free energy   A thermodynamic unit which combined enthalpy and entropy into a single (G); ΔG (free energy) $= \Delta H$ (enthalpy) $- T$ (temperature, in K) ΔS (entropy); $\Delta G_{reaction} = \Delta G_{f\,products} - \Delta G_{f\,reactants}$.

Global warming   Is a result of increases in the amounts of carbon dioxide, methane, and nitrous oxide from natural and human-produced sources.

Glycosidic bond   A $-C-O-C-$ bond formed between two sugars in a larger molecule or polymer.

Greenhouse effect   Describes how atmospheric gases prevent heat from being released back into space, allowing it to build up in the Earth's atmosphere.

Half-life   The time needed for half the atoms of a radioactive isotope to undergo radioactive decay into a different form.

Heat capacity   Amount of energy needed to increase the temperature of a set amount of sample 1°C.

Heat of evaporation (ΔH_{evap})   Total heat needed to convert a quantity of liquid at its boiling point into vapor without a rise in temperature; also called heat of vaporization.

Heat of fusion   Amount of heat absorbed when a solid sample is converted into a liquid at its melting point without a rise in temperature; also called heat of solidification.

Heat of solidification   Amount of heat absorbed when a solid sample is converted into a liquid at its melting point without a rise in temperature; also called heat of fusion.

Heat of vaporization (ΔH_{vap})   Total heat needed to convert a quantity of liquid at its boiling point into vapor without a rise in temperature; also called heat of evaporation.

Heat transfer (Q)   Equals the amount of heat in the system as it changes during the system's work within its surroundings.

Heat of reaction   Describes heat released or absorbed during a chemical reaction, which will keep all the reactants at the same temperature; exothermic ($\Delta H < 0$) = heat ↑ (lost), temperature ↑; endothermic ($\Delta H > 0$) = heat ↓ (absorbed), temperature ↓.

Hess's law   Total energy change equals the sum of the energy changes leading to the overall reaction; $\Sigma E = \Delta A + \Delta B + \Delta C + \cdots$.

Heterogeneous reactions   Reactants present in more than one phase and rates are affected by surface area.

High-density lipoproteins (HDL)   Carry cholesterol out of the blood stream and to the liver for excretion.

Homogeneous reactions   Reactants and products are in one phase (solution).

Homologous series   Organic compounds which are very similar and react much the same way.

Hund's rule   All orbitals of a given sublevel must be occupied by a single electron before pairing begins.

Hydrocarbon   A class of organic compounds containing only hydrogen and carbon (i.e., alkanes).

Hydrolases   Catalyze hydrolysis reactions and their reverse reactions.

Hydrolysis   When polymers break down to their constituent monomers when water is added.

Hydrophilic   Polar molecules, attracted to water molecules because of polarity; interact with (dissolve in) water by forming hydrogen bonds.

Hydrophobic   Nonpolar molecules don't dissolve in water.

Hypothesis   A statement or idea that describes or attempts to explain observable information.

Ideal gas equation ($pV = nRT$)   Primarily involves concentration and gas pressure and equilibrium.

Ideal gas law   Combination of Boyle's, Charles's, Gay-Lusac's, and Avogadro's laws; $PV = nRT$.

Immiscible   Two liquids, which don't mix, but instead form separate layers, (e.g., oil and water).

International System of Units (SI)   Platinum standard measures by which all other standards are compared; SI system has seven base units from which other units are calculated.

Ionic bond   Electrons are transferred between atoms.

Ionization An atom and molecule gains or loses one or more electrons and is no longer neutral.

Ionization energy Is the energy needed to detach an electron from an atom of the element.

Island of stability Large radioactive isotopes with magic numbers of nucleons.

Isobaric process A thermodynamic procedure where pressure doesn't change.

Isomerase Catalyzes the conversion of a molecule into an isomer.

Isomers Have the same molecular formula, but different structural formulas.

Isothermal process System is kept at the same temperature during the experiment (i.e., $t_{initial} = t_{final}$).

Isotopes Are different types of atoms with the same number of protons (i.e., atomic number), but different number of neutrons (i.e., atomic mass).

Isotopes Chemically identical atoms of the same element with different numbers of neutrons and mass numbers.

Kinetic energy Energy possessed by a body because of its motion; $KE = \frac{1}{2} mv^2$.

Kinetics Describes chemical reaction rates, variable effects, atom relocation, and the creation of intermediates.

Krebs cycle A physiological cycle of enzyme-catalyzed chemical reactions, which plays a key role in oxygen use within all living cells.

Laughing gas Nitrous oxide; used by dentists to put patients to sleep during dental procedures.

Law of increased entropy Matter/energy degrades slowly over time; usable energy is permanently lost in the form of unusable energy; also called the second law of thermodynamics.

Law of mass action Explains how the rate of a specific chemical reaction is proportional to the sum of the reactant concentrations.

Law of multiple proportions States that element weights always combine in small whole number ratios.

Law, scientific A hypothesis or theory that is tested time after time with the same resulting data and thought to be without exception.

Least common multiplier (LCM)   Smallest positive number that is a multiple of two numbers.

Le Châtelier's principle   After a system at equilibrium is impacted, the initial equilibrium will change in a direction that lessens or counteracts the effect of the disturbance.

Ligase   Catalyzes reactions binding together smaller molecules into larger ones.

Linear combination of atomic orbitals (LCAO)   A quantum chemistry method for calculating molecular orbitals based on their phase relation.

Lipid   Organic compounds (e.g., fats, oils, waxes, sterols, and triglycerides) found in biological systems; insoluble in water (hydrophobic), oily to the touch, and a key ingredient within the structure of living cells.

Lipoprotein   A combination of proteins and lipids; transport lipids (e.g., cholesterol) around in the blood.

Low-density lipoprotein (LDL)   Associated with the addition of cholesterol to artery walls.

Luciferase   An enzyme that emits light when oxidizing the protein luciferin.

Lyase   Adds/removes small molecules (e.g., water or ammonia to/from a double bond).

Macromolecule   Extremely large molecule (e.g., β-carotene).

Magnetic resonance imaging (MRI)   Use of nuclear magnetic resonance of protons to create proton density images.

Matter   The basic material of which are things are made.

Metric system   A decimal-based system of measurement.

Miscible   Two or more liquids combine and form a solution (e.g., alcohol and water).

Molar heat capacity   Heat needed to raise the temperature of 1 mole of a sample by 1°C.

Molar mass   Equal to the atomic and formula masses of elements and compounds (measured in grams/mole); (also known as atomic weight).

Molarity (M)   Equals the number of moles of solute (n) per volume in liters (V) solution (i.e., concentration).

Molecular formula   Gives the exact number of different atoms of an element in a molecule.

Molecular weight   Equals the sum of the atomic weights of the atoms in a molecule of a substance.

Molecule   Composed of atoms chemically bonded by attractive forces.

Monatomic ions   Consist of one or more atoms of the same element.

Monosaccharide   A simple sugar; carbohydrate.

Nanomedicine   Medical field targeting disease or repairing damaged tissues like bone, muscle, or nerve at the molecular level; also molecular sensors for precise drug delivery.

Nanoshells   That is, gold-coated silica particles; with tunable optical properties affected by size, geometry, and composition.

Nernst equation   Energy in chemical reactions affects charge, equilibrium, and gives rise to the *cell potential* within a galvanic cell or battery.

Neutralization   Takes place when mixed acids and bases cancel each other out and become neutral (e.g., pH of 7).

Neutrons   Subatomic particles with a similar mass to their partner proton in the nucleus but with no electrical (+ or −) charge.

Nitrification   Oxidation of ammonia in wastewater to nitrite and then to nitrate by bacterial or chemical reactions.

Non–point source contamination   Non-point source effluent (e.g., storm water and agricultural runoff) enters the water supply from soils/groundwater systems runoff and the atmosphere through rainfall.

Nonpolar covalent bond   Fairly equally shared electrons between atoms in a covalent bond.

Nucleic acids   Made of nucleotide chains; nucleotides are made up of three parts: a nitrogenous heterocyclic base (i.e., a purine or a pyrimidine), a 5-carbon sugar (e.g., ribose or deoxyribose), and a phosphate group.

Ohm   SI unit of measurement for electrical resistance.

Ohm's law   Explains the amount of current flowing between two ends of a conductor (e.g., copper wire) is proportional to the potential difference between them.

Orbital diagram   Notation used to show the number of electrons in each subshell.

Organic chemistry   Chemistry of carbon and carbon-based compounds; studies matter from once-living organisms.

Oxidation   When organic compounds react with an oxidizing element (e.g., oxygen or fluorine) in high temperatures to form carbon dioxide and water.

Oxidation numbers   Based on the difference between the number of electrons an atom in the element can control and the number of electrons an atom in a compound can grab and hang on to.

Oxidizing agent   Causes the oxidation of a sample, while itself being reduced.

Oxidoreductase   Catalyzes the reduction or oxidation of a molecule.

Ozone   Natural layer of ozone molecules (O_3) located from 9 to 30 miles up into the atmosphere; blocks the Earth from the sun's damaging ultraviolet radiation.

Pauli exclusion principle   No two electrons in the same atom can be in the same configuration at the same time.

Peptide bond (amide bond)   A covalent bond formed between two molecules where a carboxyl group from one molecule reacts with the amine group of another molecule, releasing a water molecule.

pH scale   Measures the acidity of a liquid by measuring the concentration of hydrogen ions (i.e., neutral pH is equal to 7.0 on the pH scale).

Phenols   A class of organic compounds with one or more hydroxyl (–OH) groups bonded to an aromatic hydrocarbon (e.g., benzene); also called carbolic acid, benzophenol, or hydroxybenzene.

Pheromone   Compound secreted by ants and bees to attract mates, signal danger, or pass along directions to a specific location.

Phospholipid   Made up of glycerol, two fatty acids, and a phosphate group; soluble in water and oil.

Physiological saline   A 0.85% NaCl solution, given intravenously and used to stabilize patients' fluid levels in hospitals.

Point source contamination   Pollutant (e.g., bilge water, diesel leaks, factory sewage, refinery oil, and waste treatment plants) released directly into urban water supplies.

Polar covalent bond   A covalent bond where electrons are *not* shared equally between atoms.

Pollutant   A substance in the environment with a harmful impact on its surroundings and living systems.

Polyatomic ions   A molecule with an overall charge containing more than one covalently bonded atom.

Polymerization   Occurs when two smaller compounds combine to form a much bigger third compound.

Polysaccharides   Also called glycan; large high-molecular-weight molecules created when many monosaccharides combine.

Positron emission tomography (PET)   An imaging method where a computer-generated image of a biological activity within the body is created through detection of emitted gamma rays (i.e., when introduced radionuclides decay and release positrons).

Precision   Closeness of two sets of measured groups of values.

Proteins   Made of small molecules containing an amino group ($-NH_2$) and a carboxyl ($-COOH$) group.

Proton   Subatomic particle with a positive charge; with roughly 1800 times greater mass than an electron.

Radioactive tracer   A minute amount of radioactive isotope added to a chemical, biological, or physical system to study the system (e.g., barium and thallium).

Radiochemistry   Another name for nuclear chemistry.

Radioimmunology   Measures the levels of biological factors (proteins, enzymes) known to be changed by different diseases.

Rate determining step   Rate of reaction tied to the slowest reaction step; also called the rate limiting step.

Rate equation   Links the rate of a reaction to each reactant and their different orders; $R = k[A]^a[B]^b[C]^c$.

Rate laws   Describe reaction rates as dependent on reactant concentrations.

Rate limiting step   Reaction rate tied to the slowest reaction step; also called the rate determining step.

Reaction order   Connection between reactants and their concentrations to an observed rate of a reaction.

Reaction rate   (Number of impacts between reacting molecules/time) × (fraction of impacts with enough activation energy to react) × (probability factor based on orientation; probability factor brings in the shape and orientation of reacting components in a chemical reaction.

Reducing agent   Causes the reduction of a sample, while itself being oxidized.

Reference state   Defined as its most stable state at 1 bar pressure and a set temperature (e.g., 298.15 K).

Rounding   Method to drop (or leave out) nonsignificant numbers in a calculation and adjust the last number up or down.

Saturated fats   Stable hydrocarbons with the maximum number of hydrogen atoms bonded to carbons; all the carbons in the fatty acid tails of triglycerides are all singly bonded to hydrogen atoms.

Scientific method   Principles and methods used in the pursuit of knowledge involving observation, data collection, testing, and formulation of hypotheses.

Second law of electrolysis   For a certain charge, the mass of a substance at an electode is directly proportional to its weight.

Second law of thermodynamics   Matter/energy degrades slowly over time; usable energy is permanently lost in the form of unusable energy; also called the law of increased entropy.

Siemen   SI unit of measurement for electrical conductance.

Significant figures   Created to write numbers in whole units or to the highest level of confidence; number of digits written after the decimal point to measure a quantity.

Solubility   Ability of one compound to dissolve into another (i.e., measured as the ratio of grams of solute/100 grams of water at a set temperature).

Solute   An element or compound dissolved to form a solution.

Solution   A homogeneous mixture of a *solute* (the element or compound being dissolved) in a *solvent* (the solution scattering the solute molecules).

Solvation   Process of making the ions into electrolytes, which conduct electricity.

Solvent   A liquid into which an element or compound is added to form a solution.

Specific heat   Amount of heat energy needed to increase the temperature of 1 gram of a sample by 1°C.

Spin magnetism   Describes how electrons are attracted and repelled by opposite and like charges, respectively.

Standard state   Of a pure substance, mixture or solution is the basis for calculations of characteristics under altered conditions; standard temperature and pressure (STP) are equal to 25°C and 1 atmosphere (atm).

Stereoisomers   Molecules have the same atoms bonded to each other, but these atoms are arranged differently in space.

Steroid   Has a molecular structure with a central core of four joined rings; examples include estrogen (estradiol), testosterone, cholesterol, corticosteroids (i.e., cortisone), and vitamin D.

Stock system Compound naming system that includes the charge on the ion in the form of a Roman numeral [e.g., ferric chloride ($FeCl_3$) is iron(III) chloride].

Structural formula Shows how specific atoms are arranged and bonded in a compound.

Structural isomers Have the same molecular formula, but with different structural arrangements.

Surfactant A substance with a hydrophilic (water loving) head and hydrophobic (water avoiding) tail that repels water while attaching to oil.

Swamp gas Methane created by plants decaying underwater; also called marsh gas.

Temperature A measure of the intensity of heat of a substance (e.g., °F, °C, K).

Theory An explanation or model based on observation, experimentation, and reasoning, especially one that has undergone thorough testing and confirms a hypothesis.

Third law of thermodynamics As the entropy of a substance nears zero, its temperature approaches absolute *zero* (i.e., 0°K, −273.15°C, or −459.7°F).

Transferases Catalyze the movement of a group of atoms from one molecule to another.

Transuranium elements All elements with greater atomic numbers than uranium (^{92}U) (i.e., element with the highest atomic number found in nature).

Turbidity Measures the murkiness (i.e., opacity) of a solution; the murkier the solution, the higher the turbidity level.

Tyndall effect Scattering of light by colloid-sized particles in a mixture.

Universal solvent Water is known as the universal solvent.

Unsaturated fats Carbons in oils (fats) are double or triple bonded and can't accommodate four hydrogen atoms; have at least one double bond in a fatty acid chain.

van der Waals equation Modified ideal gas law which corrects for these non-ideal gas interactions.

References and Internet Sites

Akins, Peter W. *The Periodic Kingdom: A Journey into the Land of the Chemical Elements*. New York: Basic Books, 1995.

Crowe, Jonathan, et al. *Chemistry for the Biosciences: The Essential Concepts*. New York: Oxford University Press, 2006.

Ebbing, Darrell, and Gammon, Steven D. *General Chemistry*. Florence, KY: Cengage Learning, 2008.

Foglino, Paul. *Cracking the AP Chemistry Exam*, New York: Random House, 2010.

Guch, Ian. *The Complete Idiot's Guide to Chemistry*. New York: Penguin Group, 2006.

Karty, Joel, *The Nuts and Bolts of Organic Chemistry*. San Francisco, CA: Pearson Education, 2006.

Moore, John T., and Langley, Richard. *Chemistry for the Utterly Confused*. New York: McGraw-Hill, 2007.

Moore, John T., and Langley, Richard. *5 Steps to a 5—AP Chemistry*. New York: McGraw-Hill, 2008.

Pulford, I., and Flowers, H., *Environmental Chemistry at a Glance*. Malden, MA: Blackwell Publishing, Inc., 2006.

Raymond, Kenneth W. *General, Organic, and Biological Chemistry*. New York: John Wiley and Sons, 2006.

Reel, Kevin, editor. *SAT Subject Test—Chemisty*, 6th edition. Piscataway, NJ: Research and Education Association (REA), 2006.

Suggs, J. Wm., Ph.D. *Barron's College Review Series—Organic Chemistry*. Hauppauge, NY: Barron's Educational Series, Inc., 2002.

The Diagram Group. *The Facts on File Chemistry Handbook*, revised edition. New York: Facts on File, 2006.

Atmospheric Chemistry

http://www.giss.nasa.gov/research/chemistry/

http://www.esrl.noaa.gov/gmd/

www.epa.gov/climatechange/index.html

www.unfccc.de

Biochemistry

http://www.chem4kids.com/files/bio_intro.html

http://www.kidsolr.com/science/page11.html

http://www.biochemweb.org

Conversions

http://www.unit-conversion.info/

http://www.unitconversion.org/

Gas Laws

http://www.chemtutor.com/gases.htm

http://www.nclark.net/GasLaws

History of Chemistry

http://www.levity.com/alchemy/home.html

http://nobelprize.org/

http://www.fas.harvard.edu/~hsdept/

http://www.factmonster.com/encyclopedia/1chem.html

Nanotechnology

www.cnst.rice.edu

http://www.nano.gov/

Organic Chemistry

http://www.acdlabs.com/iupac/nomenclature/

http://legacyweb.chemistry.ohio-state.edu/flashcards/

Periodic Table

http://www.webelements.com

http://periodic.lanl.gov/default.htm

http://www.sciencegeek.net/tables/tables.shtml

http://www.periodicvideos.com

Pollution

http://www.epa.gov/osw/

http://www.epa.gov/p2/

http://www.umich.edu/~gs265/society/waterpollution.htm

http://www.wastenet.net.au/information/streams/organic

Radiation and Radioisotopes

http://www.eoearth.org/article/Radioisotopes_in_medicine

http://www.radiochemistry.org/nuclearmedicine/radioisotopes/01_isotopes
.shtml

http://www.epa.gov/radtown/cosmic.html

http://www.physics.isu.edu/radinf/risk.htm

http://emergency.cdc.gov/radiation/measurement.asp

Space

www.nasa.gov/home/index.html

http://nssdc.gsfc.nasa.gov/photo_gallery/

http://earth.jsc.nasa.gov/sseop/efs/

http://www.noaa.gov/satellites.html

Thermodynamics

http://www.physics4kids.com/files/thermo_intro.html

http://www.nist.gov/cstl/index.cfm

http://www.ridgecrest.ca.us/~do_while/sage/v7i1f.htm

http://ftexploring.com/energy/first-law.html

Chemistry-at-a-Glance
Study Sheets

Avogadro's number: 6.02×10^{23} atoms **1 mole** $= 6.022 \times 10^{23}$ particles/mol $=$ formula weight (grams)	**Temperature** $^\circ C = 5/9 \,(^\circ F - 32)$ $^\circ F = ^\circ C \times \tfrac{9}{5} + 32$ $K = ^\circ C + 273$	**Bond length** radius $\uparrow \rightleftarrows$ bond length $\uparrow$ bond length $\downarrow \rightleftarrows$ number of bonds $\uparrow \rightleftarrows$ bond energy $\uparrow$
Orbital diagrams - pair $\uparrow\downarrow$ (Pauli's exclusion principle) - unpaired e$^-$ have parallel spins **Hund's rule:** orbitals are singly filled ($\uparrow$) first, before doubly filled ($\uparrow\downarrow$); all electrons in singly filled orbitals $=$ same spin	**Quantum numbers** - principle (n) $=$ E level - orbital (l): s $= 0$, p $= 1$, d $= 2$, f $= 3$; $l = 0, 1, 2, \ldots (n-1)$ - magnetic (m_1): $-1 \ldots 0 \ldots +1$ - spin (m_s) $= +1/2$ or $-1/2$	**Electronegativity (χ)** $=$ ability of atom to attract electrons in a chemical bond - **ionic bond** (metal & nonmetal; large $\Delta\chi$) - **covalent bond** (nonmetal & nonmetal; small $\Delta\chi$) $+$ *nonpolar* covalent bond: $\Delta\chi = 0$; $+$ *polar* covalent bond: $\Delta\chi > 0$
Concentration - Molarity (M) $=$ moles$_{solute}$/liters$_{solution}$ (mol/L) - Molality (m) $=$ moles$_{solute}$/kilograms$_{solvent}$ (mol/kg)	**Ions** - cations: $\downarrow$electron repulsion; small - anions: $\uparrow$electron repulsion; large - isoelectronic: same number of electrons	**Charge** $=$ **(number of valence e$^-$)** $-$ **(# nonbonding e$^-$)** $-$ ½ **(number of bonding e$^-$)** - sum of formal charges $=$ overall charge
Solubility (like dissolves like) - polar/polar - nonpolar/nonpolar - polar/nonpolar $=$ *immiscible* $=$ not soluble	**Ionization Energy (I)** $=$ energy added to remove electrons - to change energy level $\rightleftarrows$ $\uparrow$ energy increase required	**Ideal gas law** ($PV = nRT$) where $R = 0.0821$ l.atm/K.mol **Combined gas law** $(P_1 V_1)/T_1 = (P_2 V_2)/T_2$ where T must be in Kelvin **Dalton's law** $P_A = P_{total}\,(n_A/n_{total})$
Limiting reagent % yield $=$ actual/theoretical $\times$ 100% v/v% $= [$(volume$_{solute}$)/(volume$_{solution}$)$] \times$ 100% **Dilution** $(M_{initial})(V_{initial}) = (M_{final})(V_{final})$ $M =$ molarity, $V =$ volume	**Dipole moment (μ):** polar > 0 nonpolar $= 0$ **Coulomb's law:** charge difference $\uparrow$, ion size $\downarrow =$ melting point $\uparrow$; solubility $\downarrow$	**Gas laws:** - Boyle's law $= P\uparrow V\downarrow$ - Charles' law $= T\uparrow V\uparrow$ - Guy-Lussac's law: $T\uparrow P\uparrow$ - Avogadro's law: $n\uparrow V\uparrow$ - Henry's law: $P\downarrow \rightleftarrows \downarrow$solubility of a gas
Electrolytes - nonelectrolyte: no ions conduct electricity - strong electrolyte (many ions conduct electricity) soluble, ionic, strong acids - weak electrolytes: weak acids	**Electron filling pattern:** 1s, 2s, 2p, 3s, 3p, 4s, 3d, 4p, 5s, 4d, 5p, 6s, 4f, 5d, 6p, 7s, 5f **Redox:** oxidation is *loss* of electrons; reduction is *gain* of electrons.	**van der Waal's equation** - $(P + n^2 a/V^2)(V - nb) = nRT$ - molecular $V\uparrow = b\uparrow$ **Standard temperature and pressure (STP)** - 1 mole gas $= 22.4$ L at $0\,^\circ C$ and 1 atm pressure

Ionic compounds (contains metal ion or ammonium ion)
- solids = ↑melting point, brittle, poor conductor of electricity
- liquid = good conductors

Metallic compounds (metal cation in an area of electrons)
- delocalized e⁻ = good conductors
- size of metal cation ↓ = melting point ↑

Molecular compounds
- ↓ melting/boiling points
- commonly soft or brittle
- poor conductors

Hydrogen bonds
- form between two molecules containing hydrogen bonded *directly* to N, O, or F
- density of ice < density of water because H-bonds form leaving empty space

Organic (carbon) compounds
<u>alkane</u> (1 bond); <u>alkene</u> (2 bonds); <u>alkynes</u> (3 bonds); <u>alcohol</u> (–ROH); **acid** (R-COOH); <u>aromatic hydrocarbon</u> (ring); <u>phenol</u> (ring + OH); <u>aldehyde</u> (R-HC=O); <u>amine</u> (R-NH₂); amide (R-2(NH₂))

Isomers = same molecular formulas, but different 3D shape

Isotope

$$^{A}_{Z}X$$

X = element symbol, Z = atomic number (number of protons), A = mass number (number of protons + number of neutrons)

Arrhenius: acid in $H_2O \rightarrow H^+$; base in $H_2O \rightarrow OH^-$

Brønsted–Lowry: acid = H+ donor
Base = H+ acceptor
- conjugate acids & bases
- amine (RNH_2) = weak base

Lewis acids & bases: acid = e⁻ pair acceptor; base = e⁻ pair donor

Buffers: $pH = pKa + log(n_{A-}/n_{HA})$

pH = a measure of acidity or basicity of a solution. pH < 7 → acidic; pH >7 → basic; pH= 7 → neutral

pH = −log[H⁺]; [H⁺] = 10⁻ᵖᴴ
- pOH = −log[OH⁻]; $10^{-14} = [H^+][OH^-]$
- pH + pOH = 14

Neutralization: acids + bases = H_2O + salt

Hydrolysis: (aq) ions as acids or bases
- cations with H⁺ = weak acid
- anions = weak bases
(*except Cl⁻, Br⁻, I⁻, NO₃⁻, ClO₄⁻, SO₄²⁻, HSO₄⁻)

Exothermic rxn. ($\Delta H < 0$) = heat ↑ (lost), temperature ↑

Endothermic rxn. ($\Delta H > 0$) = heat ↓ (absorbed), temperature ↓

1st law of thermodynamics (conservation of energy) $\Delta U = \Delta Q - \Delta W$

2nd law of thermodynamics
$\Delta S_{surr} = -\Delta H_{rxn}/T$

3rd law of thermodynamics
- as entropy of pure substances nears zero, absolute temperature nears zero

Gibbs free energy
$\Delta G_{reaction} = \Delta G_{f\,products} - \Delta G_{f\,reactants}$

Hess' Law
$\Delta H_{reaction} = \Delta H_{products} - \Delta H_{reactants}$

Calorimetry: $q = m \times c \times \Delta T(T_{final} - T_{initial})$
m = mass of sample
c = specific heat capacity
ΔT = temperature change

Kinetic energy = $1/2 \times m \times v^2$

Kinetics = rate of rxn
- rate = $\Delta[A]/\Delta time$
- Rate law = $k[A]^x[B]^y$
- 0th order (rate = k); 1st order (rate = $k[A]^1$); 2nd order (rate = $k[A]^2$)

Periodic Table of the Elements
- **Periods** = *horizontal* rows (number 1-7 on the left-hand side of the table)
- **Groups** = *vertical* columns (group members have similar properties)
[*Note:* groups across columns top (older = Roman numerals and letters; newer = 1–18]
- PT information = element name, symbol, atomic number, atomic wt, std state, group, period, color, classification, properties (e.g., density, boiling, melting, freezing points) etc.

Periodic Table

Periodic Table of the Elements

Atomic no. — 1
Symbol — H
Atomic wt. [1] — 1.0079

1. ( ) Denotes mass number of most stable known isotope

† Discovery date: 1994

1A	2A	3B	4B	5B	6B	7B	8			1B	2B	3A	4A	5A	6A	7A	8A
1 H 1.0079																	2 He 4.00250
3 Li 6.941	4 Be 9.01218											5 B 10.81	6 C 12.011	7 N 14.0067	8 O 15.9994	9 F 18.99840	10 Ne 20.179
11 Na 22.98977	12 Mg 24.305											13 Al 26.98154	14 Si 28.0855	15 P 30.97376	16 S 32.06	17 Cl 35.453	18 Ar 39.948
19 K 39.0983	20 Ca 40.08	21 Sc 44.9559	22 Ti 47.88	23 V 50.9415	24 Cr 51.996	25 Mn 54.9380	26 Fe 55.847	27 Co 58.9332	28 Ni 58.69	29 Cu 63.546	30 Zn 65.38	31 Ga 69.72	32 Ge 75.59	33 As 74.9216	34 Se 78.96	35 Br 79.904	36 Kr 83.80
37 Rb 85.4678	38 Sr 87.62	39 Y 88.9059	40 Zr 91.22	41 Nb 92.9064	42 Mo 95.94	43 Tc 98.9072	44 Ru 101.07	45 Rh 102.9055	46 Pd 106.42	47 Ag 107.868	48 Cd 112.41	49 In 114.82	50 Sn 118.69	51 Sb 121.75	52 Te 127.60	53 I 126.9045	54 Xe 131.29
55 Cs 132.9054	56 Ba 137.34	*	72 Hf 178.49	73 Ta 180.9479	74 W 183.85	75 Re 186.207	76 Os 190.2	77 Ir 192.22	78 Pt 195.08	79 Au 196.9665	80 Hg 200.59	81 Tl 204.383	82 Pb 207.2	83 Bi 208.9804	84 Po (209)	85 At (210)	86 Rn (222)
87 Fr (223)	88 Ra 226.0254	**	104 Rf (261)	105 Ha (262)	106 Sg (263)	107 Ns (262)	108 Hs (265)	109 Mt (266)	110 †	111 †							

* Lanthanide Series

57 La 138.9055	58 Ce 140.12	59 Pr 140.9077	60 Nd 144.24	61 Pm (145)	62 Sm 150.36	63 Eu 151.96	64 Gd 157.25	65 Tb 158.9254	66 Dy 162.50	67 Ho 164.9304	68 Er 167.26	69 Tm 168.9342	70 Yb 173.04	71 Lu 174.967

** Actinide Series

89 Ac 227.0278	90 Th 232.0381	91 Pa 231.0359	92 U 238.029	93 Np 237.0482	94 Pu (244)	95 Am (243)	96 Cm (247)	97 Bk (247)	98 Cf (251)	99 Es (252)	100 Fm (257)	101 Md (258)	102 No (259)	103 Lr (260)

Index

An "f" follows page numbers referencing figures; a "t", tables.

A

Absolute zero, 182
  determination, 17
Accuracy
  definition, 13
  precision, contrast, 13
Acetamide (CH_3CONH_2), 221
  formula, 221f
Acetic acid (vinegar) (CH_3CO_2H)
  butanol, combination (example), 217
  chemical structure, impact, 210
  creation, 210
  formula, 214f
Acetyl coenzyme A (acetyl-CoA), 227
Acetylene (ethyne), 199
Achirality, example, 207
Achiral molecules, 207
Acid/base contaminants, 242
Acid deposition, impact, 246
Acid rain, 245–246
  atmospheric compounds, 246
  definition, 245
Acids, 159, 160
  catalyst, usage, 210
  comparison, 164
  definition, 160
  end point, 169
  handling, care, 171
  impact, 162
  neutralization, calcium carbonate (usage), 164
  pH, 168
  safety, 170–171
  strength, 163–164
    dependence, 164
    ranking, 163t

Actinide series, 41, 43, 46
Activation energy, 184–185
  definition, 184
  free energy, overcoming, 185f
  reduction, 185
  temperature, relationship, 185–186
Addition reactions, 208
Adenine (A), 225
Adenosine triphosphate (ATP), detection, 235
Adiabatic process, 182–183
Aerosol, 97
Alchemy, 2
Alcohol
  chemical structure, 210
  organic groups, 217
Aldehydes, 204
al-Haytham, Ibn (scientific method development), 6
Alkali metals, 41, 44–45
  properties, 44, 116t
Alkaline cells, 152–153
Alkaline earth metals, 41, 45
Alkanes, 196–197
  bond types, 201
  names, examples, 201
Alkenes, 196, 197–198
  bond types, 201
Alkenyl groups, rules, 201
Alkyl groups, 201
  representation, 219
Alkynes, 196, 198–199
  bond types, 201
  triple bond, 198
  triple-bonded carbon compounds, 198f
Alkynyl groups, rules, 201
Allergy, suffering, 200f
Allotropes, definition, 60

Alloys, 48–49
    creation, 49t
    definition, 48
Alpha particles, 258, 263–264
    decay, occurrence, 263
    loss, 263–264
    radiation type, 262
    strength, 264
Aluminum
    Alzheimer's disease, relationship, 246
    conductor, 154
    deposition, 246
    reducing agent, 147
Alzheimer's disease, aluminum (relationship), 246
Amalgam, 49
Americium
    Am-241, usage, 275
    Am-243, creation, 262
    inner transition metals, 46
Amides, 220–221
    formation, 220
    naming, 221
Amines, 219–220
    classes, 219–220
    naming, 219
    primary amines, 219
    secondary amines, 220
    tertiary amines, 220
Amino acids, 223–224
    functional groups, 223f
Amino group (-NH$_2$)
    hydrocarbon group, bonding, 219
    location, 219
Ammonia (NH$_3$)
    chemical formula, 28
    formation, 167
    ion, formation, 137
Ammonium carbonate (NH$_4$)CO$_3$
    ions, balancing, 138
    usage, 194
Ammonium chloride (NH$_4$Cl), electrolyte, 152
Ammonium ion (NH$_4$), formation, 137
Ammonium ion (NH$_4$1$^+$), polyatomic ion, 135
Amorphous solids, 56–57
Amphiprotic ions/molecules, 163
Amphoteric ions/molecules, 163
Anabolism, 231
Analytical chemistry, 2
Anions
    formation, 134
    interaction, 138
Anode, 150
Applied science, 4, 6
    NIH/NASA support, 4
Aquated (solution), 144

Aristotle, four-element theory, 2
Aromatic hydrocarbons, 199–202
    bonding, 222
Arrhenius, Svante, 160, 184, 186
Arrhenius theory, 160–161
Arsenite (AsO$_3$$^{3-}$), negative charge, 135
Asphalt, resistance, 154
Atherosclerosis, 234
    biomarkers, 235
Atmosphere, 79–80
    methane, interaction, 251
    representation, 82
Atmospheric methane, increase (causes), 251
Atmospheric ozone, CFC/nitrogen oxide
        interaction, 253
Atmospheric pressure, 81–82
    definition, 81
Atomic bomb, creation, 266
Atomic mass, superscript, 259
Atomic number, 36–37
    definition, 37
    electronegativity values, comparison, 129f
    identification, examples, 260
    subscript, 259
    writing, example, 260
Atomic number, example, 25
Atomic orbitals, linear combination, 121
Atomic structure, 21
    parts, 22–24
    quiz, 30–31
Atomic theory, 21
    beginnings, 22–26
    quiz, 30–31
Atomic weight, 37–41
    interactions, 39
    measurement, 98–99
    plotting, 37
    research result (Meyer), 37f
    role, 39
Atoms
    bonding, 29f
    bonds, alternation, 121–122
    characteristics, 22
    Democritus, investigation, 22
    electrons, sharing, 202
    energy levels, change, 116
    Greek prefixes, numbering, 134t
    orbital configuration, 112
    presence, 28t
    weak bonds, formation, 164
Aufbau principle, 112–113
    definition, 112
Avogadro, Amedeo, 87
Avogadro's law, 87–88
    definition, 88

B

Baking soda (sodium bicarbonate) ($NaHCO_3$), 138
  base, action, 161
Barium
  Ba-37, diagnostic usefulness, 274
  Ba-136, medical usage, 274
  characteristics, 60
  location (IIA), 45
  radionuclide, usage, 274
Barium (Ba^{2+}), cation (example), 135
Barometer, 81–82
  indication, 81f
Bases, 159, 160
  definition, 160
  end point, 169
  handling, care, 171
  impact, 162
  pH, 168
  proton absorption, 164
  safety, 170–171
  strength, 164–165
Basic science, 4, 6
  NIH/NASA support, 4
Becquerel, Antoine, 258
Becquerel (Bq), radiation amount, 265–266
Beer, brewing, 210
Benzene (C_6H_6)
  aromatic hydrocarbon, 199–200
  bonds, formation, 122f
  formula, 214f
  ring, 222
Benzenol
  formula, 222f
  IUPAC name, 223
Beryllium, energy levels, 119f
Berzelius, Jons Jacob, 204
Beta-carotene, structure, 215f
Beta particles, 258, 264–265
  radiation type, 262
Bicarbonate (HCO_3^-), formation, 138
Binary covalent compound, 130–131
  IUPAC naming method, 133
Bioavailability, 237–238
  definition, 238
Biochemical molecules, variety, 214
Biochemical reactions (reduction), enzymes
    (impact), 226
Biochemistry, 2, 213
  characteristic, 214
  quiz, 239–240
Biodegradable (term), 253–254
  definition, 253
Biological activity, alteration, 189
Biological enzymes, 226

Biological markers (biomarkers), 234–236
  definition, 235
Biological molecules
  example, 214f
  ring structures, 200
  types, 214
Biological oxygen demand (BOD)
  definition, 243
  environment test, 243
Biological risk, measurement, 272
Bioluminescence, 235
Biomarkers (biological markers), 234–235
Bismuth (Bi), isotopes, 262
Body
  delivery systems, 237
  functions, elements (importance), 39t
Bohr, Niels, 22
Boiling point, 65, 71–72
  definition, 71
  range, 72t
Bonded atoms, electronegativity, 130–131
Bonding, 57–60
  angle, electronegativity, 130–131
  pairs, nonbonding pair replacement, 124
Bonding, occurrence, 28
Bond polarity, 164, 202–203
  definition, 202
Book of Optics (Al-Haytham), 6
Boric acid, weak acid, 163
Boron, electron configuration (example), 117
Boron trifluoride (BF_3), IUPAC naming, 133
Boyle, Robert, 83, 160
Boyle's law, 83–84
  equation, 83
  example, 83–84
British Alkali Acts Administration, 245
Bromate (BrO_3^-), 140
Bromine (Br)
  addition, example, 208
  atomic number, 129
  characteristic, 44
  extraction, 144
  methane, reaction, 208–209
Bromite (BrO_2^-), 140
Brønsted, Johannes, 161
Brønsted-Lowry acids, 161
  definition, Arrhenius definition (comparison),
    161
Brønsted-Lowry bases, 161
Brønsted-Lowry reaction, example, 161
Buffering capacity, 170
Buffers, 169–170
  definition, 169
Butane (C_4H_{10}), purification/collection, 72
Butanol, acetic acid (combination), 217

C

Cadmium, environment test, 243
Cadmium sulfide, shell, 47
Caffeine, formula, 214f
Calcium
    electron configuration, 116
    extraction, 144
Calcium carbonate ($CaCO_3$), acid neutralization, 165
Calcium chloride (CaCl)
    electrical magnetism, 58
    linear bond, 123
Calcium nitride [$Ca(N_3)_2$], 139
Calorie, 178
Calorimetry, 180–181
    definition, 180
Carbamide, 221
    formula, 221f
Carbohydrates, 214, 229–231
    definition, 229
    functions, 230f
    lipids, comparison, 231
Carbon
    atomic number, 115
    atoms
        continuous link, interruption, 218
        single bonds, formation, 196
    chemistry, 194
    compounds, study (importance), 132
    cycle, industrial component, 252f
    dating, 270–271
    double bonds, location, 202f
    electronegativity, 131
    family, comparison, 195
    functional groups, 203–204
        addition, 215t
        variety, 203
    groups
        attachment, 234f
        functional groups, addition, 204
    importance, 194–196
    isotope (C-14), creation, 170
    orbitals, 194
    quiz, 211–212
    rotation, increase, 197f
    storage, 252t
Carbon-based compounds, 194
Carbon-based molecules, bond types, 123f
Carbon-carbon bonds, 194
Carbon-carbon bonds, rotation, 197f
Carbon dioxide (CO_2), 79–80
    absorption, 252
    concentration, trends, 248f
    gases, origin, 249
    greenhouse gas, 248–249

Carbon dioxide (CO_2) (*Cont.*):
    linear bond, 123
    oxidization, 244
    standard heat of formation, 179
Carbon hydrates, 229
Carbon-hydrogen bonds, 194
Carbonic acid (H_2CO_3)
    formation, 245
    proton loss, 138
Carbon monoxide (CO), reaction, 148
Carbons
    prefixes, 201
Carbon tetrachloride (CCl_4)
    dipole moment, 132
    geometry, 124
    polar bonds, 203
Carbonyl atom, 218
    nitrogen, attachment, 221
Carboxyl groups (COOH), 228
Carboxylic acid (R-CO-OH), 204, 216–217
    bonding, 216
    mixture, 210
    naming, 217
    organic groups, 217
Carlisle, Anthony, 153
Catabolism, 231
Catalyst, definition, 185
Cathode, 150
Cathode ray tube, usage, 23
Cations
    charges, balance, 151
    formation, 134
    interaction, 138
Cavendish, Henry, 166
Cell membranes, hydrophilic portion (direction), 234f
Cellulose, 229
    formula, 230f
Celsius, Fahrenheit/Kelvin conversion, 16–17
    example, 17
Central (body) cubic solid, 57f
Cesium
    characteristic, 44–45
    configuration, example, 117
    solid/liquid transformation, 64
Chalcogens, 41
Charged ions, formation, 134
Charles, Jacque, 84
Charles's law, 84–85
    definition, 84
    example, 85
Chemical atomic weights, research, 204
Chemical bases, familiarity, 164
Chemical bonds, 127
    quiz, 141–142

Chemical burns, 170
Chemical formulas, examples, 27–28
Chemical kinetics, 183–188
    definition, 183
    experimental conditions, 184
Chemical naming, clues, 140f
Chemical nomenclature, 35–36
    definition, 36
Chemical oxygen demand (COD), environment test,
        243, 244
Chemical properties, 60
Chemical reaction rates, impact, 184
Chemicals, nicknames, 36
Chemistry
    definition, 2
    disciplines, 2, 4
        list, 3t
    quiz, 18–19
    SI base units, usage, 9t
Chernobyl nuclear power plant accident, 266
Chirality, 206–207
    example, 207
    factors, 207f
Chloride (Cl^-), anion (example), 135
Chlorine
    atoms, placement, 206f
    extraction, 144
    methane, reaction, 208–209
Chlorobromoiodomethane, non-superimposable isomers,
        206
Chloroethane, yield (example), 208
Chlorofluorocarbons (CFCs), atmospheric ozone
        interaction, 253
Chloroform ($CHCl_3$)
    dipole moment, 132
    polar bond, 132–133
Chlorophyll, 229
Cholesterol, 233
    carbon groups, attachment, 234f
Chromium
    deposition, 246
    oxidation state, 149
Chromophores, 235–236
    bonding ability, 235
    definition, 235
Cigarettes, nitrogen oxide production,
        250
Cinnamaldehyde (C_9H_8O), 200
Cinnamon, oil (C_9H_8O), formula, 200f
cis-1,2-dichloroethene, formula, 206f
cis–trans isomerism, 206
cis–trans isomers, 206
    formulas, 206f
Citric acid, presence, 216
Coal tar, 222

Cobalt
    cobalt-60, radionuclide, 274
    sulfur, combination, 51
Codons, 225
Coenzymes, 227
Cofactor binding, 226
Cofactors, 227
Colloids, 97
    mixture, parts, 97
    phases, 97
Combined gas law, 86–87
Common naming system, 139–140
Compounds, 64
    chemical properties, impact, 71
    molar mass, determination, 99
    naming
        example, 133
        guidelines, 140
    structural formulas, 29f
Concentration, change, 103–105
Condensation, 65, 68–69
    definition, 68
    dynamic equilibrium, 69f
    rate, change, 68
    reaction, 224, 231
Conductors, 154–155
Conjugate acid, definition, 163
Conjugate acid-base pairs, 162–166
    importance, 163
Conjugate base
    creation, 162
    definition, 163
Conjugate strength, 164
Contaminants
    presence, 242t
    types, 242
Contamination, 242–245
    release, methods, 242
Continuous colloid phase, 97
Conversion factors, 15–17
Copper, 45–46
    characteristic, 46
    conductor, 154
    density, 61
    deposition, 246
    environment test, 243
    ionic states, 135
    percentage, 49
Copper carbonate/sulfate, 60
corticosteroids/cortisone, 233
Coulomb, 154
Coulomb-meters (C-m), 133
Counted significant figures, 11
Counting atoms, Greek prefixes,
        134t

Covalent bond, 26
  characteristic, 128–130
  definition, 128
  formation, 224
  nonpolar covalent bond, 131
  polar covalent bond, 131
Covalent compounds, 130–131
  naming, 133–134
Covalent pair, 131
Covalent solids, 59–60
Crabs, chitin shells, 224
Cracking, 209
Creosote, 222
Crutzen, Paul, 253
Crystalline solids, 57
  atomic arrangements, 57f
Crystallization, 57–60
Cuprous sulfate, usage, 139
Curie, Marie/Pierre, 258, 265
Curie (Ci), radiation amount, 265–266
Curium
  Cm-247, creation, 262
  power supply, 275
Curran, Samuel, 266
Current
  changes, measurement, 168
  flow, 155
Cyclohexane (C_6H_{12}), 199
Cytosine (C), 225

D

Dalton, John, 90
Dalton's law (partial pressures), 90–91
  equation, 90
  example, 92
Daughter nuclide, 261
de Broglie wavelength, 48
Debye units (D), 133
Decane ($C_{10}H_{22}$), 194
Decay rate, 261
Defined significant digits, 11
Deforestation, nitrogen oxide production, 250
Dehydration
  reaction, 230
  synthesis, 224
Delta (Δ), 177
Democritus, atomos (term), 22
Density, impact, 61
Deoxyribonucleic acid (DNA), 225
  diameter, range, 238
  double helix structure, 225f
  double stranding, 224
De Rerun Natura (Nature of Things), 22
Designer proteins, research, 225

Detectors, types, 266–267
Detergents, 216
Deuterium, 259
Diamond, formation, 60
Diatomic chloride, addition, 208
Diatomic compounds, 131
Dichloroethane, cis/trans bonding, 206f
Die Modernen Theorien der Chemie (Modern Theory of
    Chemistry) (Meyer), 37
Diethyl ether, formula, 195f
Diffusion, 78
Dimensional analysis (factor-label method), 15–16
Dimer, 230
Dimethylacetylene, triple bonds, 198
Dipole imbalance, 133
Dipole moment, 132–133
  definition, 132
Direct (point source) contamination, 242
Disaccharides, 229, 230
Dispersed colloid phase, 97
Dissolution, 246
  definition, 245
Dissolved oxygen (DO)
  definition, 243
  environment test, 243
Double bonded carbons, examples, 197f
Double-bonded hydrocarbon, addition, 208
Double bonds, 123f
Double-stranded nucleic acid, composition, 224
Dow, Herbert Henry, 144
Dow Chemical Company, 144
Drug-delivery methods, creation, 238
Dry cells, 152–153
D-type orbitals, 110
Dynamic equilibrium, 69

E

Electrical resistance, dependence (absence), 155
Electricity
  electron movement, 151
  passage, 153
Electrochemical cell, 149–153
  electrolytic cell, contrast, 153
Electrochemical reactions, study (Faraday), 153
Electrochemistry, 143
  introduction, 144
  quiz, 156–157
Electrode, oxidation/reduction (occurrence), 150
Electrolysis, 153–154
  definition, 153
  first law, 154
  laws (Faraday), 153–154
  methods, creation, 144
  second law, 154

Electrolyte, definition, 59
Electrolytic cell, 153
  electrochemical cell, contrast, 153
Electronegativity, 128–129
  definition, 128
  Pauling scale, 130t
  values, atomic number (comparison), 129f
Electron pairs
  linear bond, 123
  octahedral geometry, 124
  tetrahedral geometry, 124
  trigonal bipyramidal geometry, 124
  trigonal planar geometry, 124
Electrons, 22, 23–24
  arrangements, 128
  attraction/repulsion, 111f
  capture, 265
  characteristics, 26t
  configuration, 110
  definition, 23
  energy levels, 110–115
    capacities, 111t
  gain (reduction), 144
  loss (oxidation), 144
  molecular orbital, allowance, 121
  movement, 151
  negative charge, 23
  number, transfer, 154
  plum-pudding model, 24f
Electropositivity, 128–129
  definition, 128
Electrostatic interactions, 138
Elements, 33
  bonding determination, 110f
  bonding potential, 130t
  classes, 41
  decay rate, differences, 262
  definition, 34
  identification (Lavoisier), 34
  interactions, deciphering, 133
  languages, 35t
  molar mass, 98–99
  names
    letter/symbol, 36
    writing, 259
  naming, examples, 260
  number (Periodic Table), 40
  number (increase), periods (impact), 42
  placement, 118
  properties, differences, 118
  quiz, 53–54
  reaction, 184
  reactivity, importance, 118–119
Empirical gas laws, 83–87
Emulsification, assistance, 232

Emulsifying agent, definition, 233
Emulsion, 97
Enantiomers, 206–207
Endothermic reaction, 178, 179
  temperature, increase, 189
Energy
  changes, sum, 179
  gain/loss, 179
  wavelengths, 237f
Energy orbitals, parallel spin, 121
English system, U.S. usage, 11
English units, metric units (comparison), 11t
Enhanced permeability and retention (EPR) effect, 236
Enthalpy, 178
  change (ΔH), 181
  temperature/product, sum, 183
Entropy
  definition, 181
  increase, law, 181
  representation (joules per kilogram), 183
Environmental chemistry, 241
  characteristic, 242
  quiz, 255–256
Environmental markers, 243
Environmental pH, 244–245
Enzyme-catalyzed chemical reactions, 228
  example, 228f
Enzymes, 226–228
  active site, arrangement, 226
  importance, 226
  specificity, 226
  substrates, connection, 227
Equations, reaction orders, 186t
Equilibrium, 188
  system
    pressure, increase, 189
    volume, condensation, 189
Equilibrium constant (K_{eq}), 188
Equivalence point, discovery, 169
Esterification, 210
  reaction rate, hydrochloric acid (impact), 210
Esters, 204, 217–218
  chemical formulas, 217
  naming, 218
  smells, 218t
Estrogen (estradiol), 233
Ethane (C_2H_6), alkane, 196
Ethanol (C_2H_5OH)
  creation, 210
  equation, 187
  heat capacity, 180
  molecular formula, 27
Ethene (C_2H_4)
  addition, 208
  diatomic chloride, addition, 208

Ethene (C₂H₄) (Cont.):
  double bond, formula, 197f
  double bonded carbon molecule, 198
  hydrogen chloride, addition (example), 208
  molecules, electrons (sharing), 198f
  usage, 197
Ethyene (acetylene), 199
Ethyl acetate (CH₃COOOC₂H₅), 187
Ethylacetylene (1-butene), formula, 198f
Ethylene (C₂H₄), usage, 197
Ethylene glycol (CH₂OHCH₂OH), water (mixture), 73
Ethyl ethanoate, reaction, 218
Ethyne, bromine (addition), 208
Evaporation, 65, 70–71
  heat, 179
  rate, change, 68
Exothermic reaction, 178, 179
  temperature, increase, 189
Experiment
  definition, 7
  usage, 7–8
Experimental results, explanation/prediction, 186
Experimentation, scientific method step, 6
Exponential notation, 10t

F

Face-centered cubic solid, 57f
Factor-label method (dimensional analysis), 15–17
Fahrenheit, Celsius/Kelvin conversion, 16–17
  example, 17
Faraday, Michael, 153
Fats, 231, 232
Fat-soluble vitamins, energy/antioxidant source, 232
Fatty acids, 232
  hydrophobic hydrocarbon chain/hydrophilic carboxyl
    group, presence, 232f
  saturation, 232–233
  types, 233
Fermentation, 210
Ferric chloride (FeCl₃), 140
Ferrous chloride (FeCl₂), 140
First-order reaction equations, example, 187
Fluid dynamics, 97
Fluorescence, 48, 235
Fluorine, methane (reaction), 208–209
Fluoromethane (CH₃F), potassium bromide (substitution
    reaction), 209
Fluorophores, 235
Foods, acid/base, 166t
Formation, standard heat (ΔH°), 178–179
Forward reaction rate
  division, 188
  reverse reaction, ratio, 188
Fossil fuels, burning, 250

4-chlorobenzenol, formula, 222f
Four-element theory (Aristotle), 2
Fractionation, 72
Free energy (G), 182–183
  difference, 183
  impact, 185f
Freezing point, 72–73
Fructose (C₆H₁₂O₆), 229
  molecular formula, 27
Fruits, ester (smells), 218t
F-type orbitals, 110
Functional groups, 199–200, 203–204
  addition, 215t
    example, 204
Fusion, heat, 179–180

G

Gadolinium oxide, usage, 138
Gadolinium oxide (Ga₂O₃), formula, 139
Galileo, scientific method development, 6
Gallium (Ga-67), medical usage, 274
Galvani, Luigi, 150
Gamma rays, 265
  radiation type, 262
Gases, 77
  atmosphere, 79–80
  atmospheric pressure, 81–82
  characteristics, 78–79
  combined law, 86–87
  definition, 78
  diffusion, 78
  empirical laws, 83–87
  Graham's law, 80
  gravity, impact, 81
  ideal law, 88–90
  laws, combination, 88
  matter, 79t
  nonmetal form, 51f
  quiz, 92–93
  standards, 88
  standard temperature/pressure units, usage, 89t
  theory, kinetic theory, 80
  volcanic release, 249t
  work, 177
Gay-Lussac, Joseph, 85
Gay-Lussac's law, 85–86
  definition, 85
  example, 86
Geiger, Hans, 24, 266
Geiger counter, 266
General Conference on Weights and Measures,
  SI adoption, 9
Genetic blueprint, 224
Geometric isomers, 206

Gerlach, Walter, 111
Gibbs, Josiah, 182
Gibbs free energy, 182–183
Glass, resistance, 154
Global warming, 248–251
   consequences, 248
Glucose, 229
   conversion, 231
   formula, 230f
   monosaccharide, 231
Glycans, 230
Glycerol, 232
Glycogen, 230, 231
Glycosidic bonds, 230–231
Gold (Au), 45–46
   characteristic, 45
   conductor, 154
   ionic states, 135
   percentage, 49
Gold nuclei, collision (rarity), 24–25
Graham's law, 80
   equation, 80
Gram per liters-second, 183
Gravity, force, 82
Greek Academy, 2
Greek prefixes, 134t
   usage, examples, 52
Greenhouse effect, 247
Greenhouse gases, impact, 247, 247f
Ground state, 116
Groups, 42, 43
   definition, 43
   display, 195
   numbering plan, IUPAC setup, 43
   presence, 43e
Guanine (G), 225
Guldberg, Cato Maximillian, 188

H

Half-life, 267–271
   definition, 268
Half-reactions, 144
   combination, 145
Halogens, 41, 50–51
   electronegative element example, 146
   group, methane reaction, 208–209
   salt formers, 50
   substitution reactions, example, 209
Halomethanes, creation, 209
Handedness, 206–207
Heat
   absorption, 178f
   capacity, 180
   generation, amount (joules), 181

Heat (*Cont.*):
   intensity, measure, 16
   potential energy/work, relationship, 177
   reaction, 179
   trapping, greenhouse gases (impact), 247t
Heat of evaporation, 179
Heat of formation, 183
Heat of fusion, 179–180
Heat of solidification, 179–180
Heat of vaporization, 179
Heavy metals, environment test, 243
Heavy water, creation, 259
Heisenberg, Werner, 122
Heisenberg's uncertainty principle, 122
Hematite, 46
Hemoglobin, 224
Hess's law, 179
Heterogenous mixtures, 63
High-density lipoproteins (HDLs), 234
Histidine, formula, 200
Homogenous materials, phase, 63
Homogenous mixtures, 63
Hormones, 224
Household solutions, acid/base, 166t
Hund, Friedrich Hermann, 114, 120
Hund's rule, 114–115, 121
   definition, 114
   example, 115
Hybridization, 120
Hydrocarbons, 196–200
   contrast, 215–216
   cracking, example, 209
   group, amino group (bonding), 219
   IUPAC naming method, 200–201
   substitution reactions, example, 209
   tails, 233
   triple bonds, 198
Hydrochloric acid (HCl)
   dissolving, 160
   encounter, 171
   examples, 104
   impact, 210
   sodium hydroxide, mixture (example), 162
Hydrogen (H_2)
   atom, carboxylic acid (bonding), 216
   bonds, strength, 67
   compounds, 166–167
   covalent bond, example, 131
   diatomic compounds, 131
   electronegativity, 131
   electrons, loss, 136–137
   forms, 258–259, 259f
   heating, 147
   importance, 166–167
   nitrogen, combination, 167

Hydrogen (H₂) (Cont.):
  oxygen, combination, 166
  reducing agent, example, 148
Hydrogen chloride, addition (example), 208
Hydrogen cyanide, exposure, 102
Hydrogen halides, creation, 209
Hydrogen ions (H⁺), 136–138, 167
  concentration (increase), acids (impact), 167–168
  proton, status, 161
  release, 160
Hydrogen peroxide, chemical reaction, 226
Hydrogen sulfide (H₂S), 167
Hydrolases, 227
Hydrolysis, 231
Hydronium ion (H₃O⁺), polyatomic cation, 135
Hydrophilic carboxyl group, presence, 232f
Hydrophilic hydrocarbons, hydrophobic hydrocarbons
    (contrast), 215–216
Hydrophilic (polar) molecules, 216f
Hydrophilic tail, 216
Hydrophobic hydrocarbon chain, presence, 232f
Hydrophobic hydrocarbons, hydrophilic hydrocarbons
    (contrast), 215–216
Hydrophobic (non polar) molecule, 216f
Hydrophobic tail, 216
Hydroxide ion (OH⁻)
  making, 137
  release, 160
Hydroxonium ion (H₃O⁺), 167
Hydroxybenzene, 222
Hydroxyl group, attachment, 222f
Hypobromite (BrO⁻), 140
Hypothesis
  creation (Lavoisier), 7
  definition, 7
  scientific method steps, 6, 7

I

Ideal gas law, 78, 88–90
  modification, 90
Imidazole (C₃H₄N₂)
  formula, 214f
  organic compounds, 200
Incinerators, nitrogen oxide production, 250
Increased entropy, law, 181
Indirect (non-point source) contamination, 242, 243
Inert gases (noble gases) (group VIII), 41
Inner transition metals (lanthanide/actinide series), 41, 46
Inorganic arsenic, ingestion (problems), 102
Inorganic chemistry, 2
Inorganic contaminants, 242
Inorganic polyatomic ions, negative charges, 135
Inorganic salt, usage, 194
Insulators, 154–155

Internal energy (U), 176
International System of Units (SI), 9
  base units, 9t
  Sievert (Sv), usage, 272
International Union of Pure and Applied Chemistry
    (IUPAC)
  binary covalent compound naming method, 133
  hydrocarbon naming method, 200–201
  naming system, Stock system (usage), 140
  numbering plan setup, 43
Iodine (I-123), medical usage, 274
Ionic bonds, 26, 138–140
  formation, 128, 138
Ionic compounds
  balancing, 138–139
  naming, 139–140
Ionic equilibrium, importance, 155
Ionic solids, 58–59
Ionization, 134
Ionization energy, 118–119
  definition, 118
Ionizing radiation, impact, 271
Ions, 136–138
  characteristics, 134–138
  formation, 134
  monatomic ions, 135
  types, 135–136
Iron (Fe), 46
  alloys, creation, 49
  ionic states, 135
Island of stability, 267
Isomerases, 227
Isomeric shapes, impact, 205
Isomers, 204–208
Iso-pentane, formula, 205f
Isothermal, isobaric thermodynamic system, potential work
    (availability), 182
Isothermal process, 182–183
Isotopes, 258–260
  definition, 258
  naming, 259–260

J

Joule, determination (example), 181

K

Kelvin
  Celsius/Fahrenheit conversion, 16–17
    example, 17
  scale, absolute zero, 182
Kerotakis device, usage, 17f
Ketones, 204
Kilocalories per mole, 179

Kilojoules per mole, 178, 183
Kinesis, 183
Kinetic energy, 176
Kinetic theory, 80
  equation, 80, 84
Krebs, Hans, 228
Krebs cycle, 227, 228
  enzyme-catalyzed chemical reactions, 228f

L

Lactate, creation, 210
Lactose, 230
Lanthanide series, 41, 43, 46
Lard (solid animal fat), 232
Lavoisier, Antonie, 6–7
  elements, identification, 34
  hypothesis, creation, 7
Law, definition, 8
Law of mass action, 188
Lead (Pb)
  deposition, 246
  environment test, 243
  Pb-82, 267
Lead acid battery (wet electrochemical cell), 151
Leaf burning, nitrogen oxide production, 250
Least common multiplier (LCM), 138–139
Le Châtelier's principle, 188–189
  definition, 189
  prediction, 189
Leclanché, Georges, 152
Lemons, citric acid (presence), 216
Lewis, G.N., 119
Lewis electron-dot structures, 120
Ligases, 227
Light
  measurement, 236
  wavelengths, 237f
Lightning, nitrogen oxide production, 250
Limestone ($CaCO_3$) dissolution, 246
Linear combination of atomic orbitals (LCAO), 122
Lipids, 214, 231–234
  carbohydrates, comparison, 231
  proteins, combination, 233–234
Lipoproteins, 233–234
Liquids, 55
  boiling point, 71–72
    range, 72t
  bonds, strength, 67t
  characteristics, 64–73
  condensation, 68–69
  density, 64–65
  dynamic equilibrium, 69
  evaporation, 70–71
  freezing point, 72–73

Liquids (*Cont.*):
  immiscible characteristic, 97
  impurity, 71
  molecules
    escape, 67
    forces, 66
  nonmetal form, 51f
  quiz, 74–75
  states, 65t
  surface tension, 65–66
  vaporization, 67–68
  viscosity, 66–67
Lithium dichromate ($Li_2Cr_2O_7$), oxidation state
    (determination), 149
Litmus paper
  creation (Boyle), 160
  function, 167
Living organisms, carbohydrates functions, 230f
Living systems, chemistry, 214
Low-density lipoproteins (LDLs), 234
  levels, elevation, 235
Lowry, Thomas, 161
Luciferase, usage, 235
Luminescent tags, luciferase (usage), 235
Lyases, 227

M

Macromolecules, 214
  branching, 194
Magic numbers, 267
Magnesium
  extraction, 144
  orbital configuration, 113f
  reducing agent, 147
Magnesium (Mg^{2+}), cation (example), 135
Magnesium chloride ($MgCl_2$)
  combination, 138
  naming, 139
Magnetic resonance imaging (MRI), 138
Main groups, 43
Maltose, 230
Manganese, reducing agent, 147
Manganese dioxide (MnO_2), 152
Manhattan Project, 266
Manometer, 82
Marsh gas (methane), 251
Mass, mole conversion, 99
Mass action, law, 188
Mass-by-mass percent, 100
Mass-by-volume percent, 100
Matter
  characteristics, 2, 34
  conservation, law, 177
  disorder, 181

Matter (*Cont.*):
  divisions, 63
  forms, 34, 56f
  properties, 34
Measurements, 8–13
  calculation, precision, 15
  decimal system
    establishment (French Academy of Sciences), 8–9
    suggestion (Mouton), 8–9
  multiplying/dividing, 14
Melting point, 39
Mendeleyev, Dimitri, 38, 40, 41
Mercury
  characteristic, 44
  environment test, 243
  impact, 49
  reducing agent, 147
Metabolic reactions, acceleration, 224
Metabolism, 214
Metal alloy
  creation, 49
  example, 49
Metal ions, structures, 47
Metallic bond, 58
Metallic crystals, 47–48
Metallic solids, 58
Metalloids, 50
  definition, 50
  presence, 44e
Metallurgy, 43–52
Metal oxide crystals, 47
Metals
  characteristics, 44
  chemistry, 43–52
  creation, 49
  electrons, loss, 134
  mixture, 49t
  naming, 51
  nonmetals, contrast, 43–50
Methane (CH_4), 250–251
  alkane, 196
  atmosphere, interaction, 251
  bond angles, display, 197f
  levels, increase, 251
  origin, 251
  purification/collection, 72
Methanoic acid, creation (example), 217
Methanol (CH_3OH), chemical structure, 210
Methylacetylene (propyne), formula, 198f
Methyl alcohol (CH_3OH), formation (example), 148
Metric system, 8–9
Metric units, preference, 11t
Meyer, Lothar, 37–41
  periodic table, updating, 39
  research results, 37f

Milk, lactic acid (presence), 216
Miscibility, 97
Miscible, term (usage), 97
Mixtures, 63
  heterogenous mixtures, 63
  homogenous mixtures, 63
m-Nitrophenol, formula, 222f
Modern chemistry, characteristics, 2, 4
Modern Theory of Chemistry (*Die Modernen Theorien der Chemie*) (Meyer), 37
Molar gas density/volume, 89
Molar heat capacity, 180
Molarity (moles per liter), 100
  definition, 100
Molar mass (MM), 89, 98–99
Molar volume, equation, 89
Mole
  characteristic, 98
  definition, 98
  example, 98
  mass conversion, 99
  temperature, rise, 180
Molecular compound, 26
Molecular formulas, 27–28
  atoms, presence, 28t
  definition, 28
Molecular geometry, 123–124
Molecular markers (signature markers), 234
Molecular orbitals, filling, 121
Molecular orbital theory, 120–121
  valence bond theory, contrast, 120–121
Molecular oxygen (O_2), oxidant (usage), 147
Molecular solids, 57–58
  intermolecular forces, 58
Molecular weight (MW), 99
Molecules, 26–29
  condensation, 231
  definition, 27
  double bonds, impact, 198
  formation, 26
  potential energy, 121
  research, 225
  shape/geometry, importance, 123
Mole per liters-second, 183
Molina, Mario, 253
Molybdenum (Mo), electronegativity values (example), 129
Monatomic anions, tracking, 137t
Monatomic cations, tracking, 136t
Monatomic ions, 135
Monomers, 231
Monosaccharides, 229
Monounsaturated fats, 233
Motion, energy, 176
Mouton, Gabriel (decimal measurement system suggestion), 8–9

Muller, E.W., 266
Mulliken, Robert, 110
Mullis, Kary, 225

N

Nanocrystals, 47
Nanomedicine, 236–238
   definition, 236
Nanoparticles, 236
Nanoshells
   creation/tuning, 237f
   definition, 236
National Aeronautical and Space Administration (NASA)
   applied/basic science support, 4
   spinoffs, 5t
National Institutes of Health (NIH), applied/basic science
      support, 4
Natural gas, methane (component), 251
Natural repositories, carbon storage, 252t
Nature, balancing act, 251–253
Nature of Things, The (*De Rerun Natura*), 22
Neo-pentane, formula, 205f
Neptunium
   decay, 262
   inner transition metals, 46
Neutralization, 161, 162
Neutral substances, pH, 168
Neutron-proton ratio, 264
Neutrons, 22
   characteristics, 26t
   definition, 26
Newlands, John, 38, 115
Nicholson, William, 153
Nickel, atomic number, 117
Nilson, Lars, 38
Nitrate esters, explosiveness, 217
Nitrates, environment test, 243
Nitric acid (HNO_3)
   change, 246
   nitrogen, atmospheric water (combination), 250
   presence, 249
Nitric oxide (NO), presence, 249
Nitride (N^{3-}), anion (example), 135
Nitrification, 244
Nitrogen
   atmospheric water, combination, 250
   cycle, importance, 250f
Nitrogen-containing compound, 221
Nitrogen dioxide (NO_2), human health impact, 101
Nitrogen-fixing bacteria, 246
Nitrogen monoxide (NO), human health impact, 101
Nitrogenous heterocyclic base, 224
Nitrogen oxide, 249–250
   atmospheric interaction, 253
   production, 250

Nitroglycerin, explosiveness, 217
Nitrous oxide (NO_2)
   laughing gas, 250
   presence, 249
Noble gases (inert gases) (group VIII), 41
Noncrystalline solids, 56–57
Nonmechanical work, ability, 183
Nonmetal polyatomic anions, existence, 139
Nonmetals, 50–51
   characteristics, 44, 50
   elements, description, 50
   metals, contrast, 43–50
   naming, 51
   presence, 51e
Non-metals, electrons (gain), 134
Non-point source (indirect) contamination, 242, 243
Nonpolar covalent bond, 131
Nonprotein ions, variety, 224
Non-superimposable isomers, 206
n-Pentane, formula, 205f
Nuclear chemistry, 257
   quiz, 276–277
Nuclear medicine, 273–274
Nuclear model, 24–26
Nuclear reactions/balance, 261
Nucleic acids, 214, 224
Nucleons, types, 22
Nucleotides, 224
Nucleus
   dense pit, 24–25
   electrons, relationship, 23f
   model (Rutherford), 24–25, 25f
Numbers, rounding rules, 14
Nutrient leaching, 246

O

Observation
   importance, 8
   scientific method step, 6
Oceans, carbon dioxide absorption, 252
Octave Rule, 38–41
Ohm, definition, 155
Ohm's law, definition, 155
Oil of cinnamon (C_9H_8O), formula, 200f
Oil of vanilia ($C_8H_8O_3$), 199–200
   formula, 200f
Oils, 231, 232
   hydrocarbon tail, double-bonded carbons, 233
Oleic acid, presence, 216
Oligopeptide, 224
Olive oil, oleic acid (presence), 216
1,2-dichloroethane, 208
1-butene (CH_2=$CHCH_2CH_3$), examination, 202
1-butyne (ethylacetylene), formula, 198f

1-methyl-2-bromobenzene, formula, 199f
1-methyl-3-bromobenzene, formula, 199f
1-methyl-4-bromobenzene, formula, 199f
1-methylbenzene, formula, 199f
*On the Relation of the Properties to the Atomic Weights
    of the Elements* (Mendeleyev), 38
Optical isomers, 207
Orbitals, 109
    characteristics, 110
    diagrams, 114
    filling, 137
        sequenced order, 113f
    hybridization, 120
    overlap, 119
    quiz, 125–126
    subshell energy levels, 110f
Organic chemistry, 4, 193
    characteristic, 194
    definition, 194
    quiz, 211–212
Organic compounds
    branched structure, 220
    formation, 215t
Organic contaminants, 242
Organic macromolecules, 223
Organic material
    carbon dating, 270
    decomposition, 222
Organic molecules, possibilities, 195f
Organic reactions, 208–210
Organics, naming, 200–202
Oxalate ion ($C_2O_4^{2-}$), organic anion, 135
Oxidation, 144–145, 210
    number, determination, 149t
    occurrence, 150
    reduction, opposite reaction, 146f
    state, 148–149
        determination, example, 149
        integers, representation, 148
Oxides, formation, 148
Oxidizing agents, 146–147
    definition, 147
Oxidizing element, organic compounds (reaction),
    210
Oxidoreductases, 227
Oxygen (O_2)
    atomic number, 129
    atoms, sulfur (addition), 139–140
    breathing, 79–80
    demand, 244
    electrons
        addition, 137–138
        configuration, example, 117
    oxidizing agent, reaction (example), 147
    solubility, 244

Ozone (O_3), 253
    depletion problem, 253

P

Parent nuclide, 261
Partial pressures, Dalton's law, 90–91
    equation, 90
    example, 92
Parts per billion (ppb), usage, 101
Parts per calculation, example, 102
Parts per million (ppm)
    measurement, 103t
    usage, 101
Parts per notation, 101–102
Pauli, Wolfgang, 113
Pauli exclusion principle, 113–114, 121
    definition, 114
Pauling, Linus, 120, 128–129
Pauling electronegativity scale, 129
    usage, 130t
p-chlorophenol, formula, 222f
Pentane
    isomer groups, 205f
    structural isomers, 205
Pentose sugar, 224
Peptide bond, 224
Peptides, formation, 224
Perbromate (BrO_4^-), 140
Percent solution, 100–103
    example, 101
Periodic chart, gaps, 38
Periodic Table, 33
    elements, number, 40
    group/period information, 119f
    quiz, 53–54
    subshells, 115–118
    tool, 40f
    updating (Meyer), 39
Periods, 42–43
    definition, 42
    impact, 42
    presence, 43e
Pesticides, environment test, 243
pH, environment pH, 244–245
pH, environment test, 243
Pharmaceuticals, creation, 238
Phase diagram, usage, 70
Phases, 63
    triple point, 70f
pH change
    minimization, 170
    steepness, 169
Phenolics, 222

Phenols, 204, 222–223
  benzene ring, 222f
  formula, 222f
  hydroxybenzene, 222
  IUPAC name, 223
  naming, 222–223
Pheromones, 217
pH meter, 168
  probe, usage, 169f
Phosphate (PO_4^{3-})
  group, 224
  negative charge, 135
Phospholipids, 231, 233
Phosphorous, environment test, 243
Photosynthesis
  cycle, 251
  products, 229
pH scale, 167–169
  formula, 168
Physical chemistry, 4
Physical properties, 60
Physics, Newton's laws (application), 189
Pi bonds, 122–123
Plant fibers, cellulose, 229
Plastic, resistance, 154
Plum-pudding model (Thomson), 23–24, 24f
Point source (direct) contamination, 242
Polar bond, increase, 164
Polar covalent bond, 131, 132
Polarity, 131–133
Polar liquids, dissolving (ability), 96
Pollutants, parts per million measurement, 103t
Polyatomic anions
  tracking, 137t
  types, 140
Polyatomic bromine anions, 140
Polyatomic cations, tracking, 136t
Polyatomic ions, 52
  definition, 135
  elements, involvement, 135
Polyethylene, 197
Polymerization, 209, 231
Polymers, breakdown, 231
Polypeptides, 224
Polysaccharides, 229, 230
Poly-unsaturated fats, 233
Positron emission tomography (PET) scan, 264–265
Positron-emitting radionuclide (tracer), 265
Potassium
  chlorine, combination, 51
  electronegativity values, example, 129
  P-40, beta particles (emission), 264
Potassium bromide (KBr), fluoromethane (substitution
    reaction), 209
Potassium hydroxide (KOH), 153

Potential energy, 176
  heat/work, relationship, 177
Potential work, availability, 182
Pounds per square inch (psi), 82
Power, current/voltage, 155
Precipitate, 62
  movement, 62f
Precipitation, acidic level, 245
Precision
  accuracy, contrast, 13
  definition, 13
Prediction, scientific method step, 6
Pressure
  calculation, pounds per square inch (psi), 82
  volume difference, multiplication, 188
Pressure, SI standard (torr), 82
Primary amines, 219
Primitive battery (voltaic pile), 150
Principal quantum number, 110
Processed cellulose, 230
Product concentrations, ratio, 188
Propane (C_3H_8), purification/collection, 72
Propyne (methylacetylene), formula, 198f
Prostate-specific antigen (PSA), 235
Proteins, 214
  catalysis/impact, 226
  characteristic, 223–225
  definition, 223
  lipids, combination, 233–234
Protium, 258
Protons, 22, 25–26
  base absorption, 164
  characteristics, 26t
  definition, 25
  donors, 161
  loss, 162
  plum-pudding model, 24f
  release, 170
P-type orbitals, 110
Purine, 224
Pyrimidine, 224

Q

Quality factor (Q), development, 272
Quantum dots, 47–48
  classification, 47
  color intensity, 48
  definition, 47
  tuning, possibility, 48

R

Rad, radiation measurement, 272
Radiation

Radiation (Cont.):
  damage, dose levels (relationship), 273t
  detection, 265–267
  detectors, types, 266–267
  dosage, 272–273
  dose size, impact, 271t
  dose unit measurement, 272t
  emission types, 262–265
  exposure, 271
    measurement, 271
  therapy (radiotherapy), 274
Radioactive contaminants, 242
Radioactive decay, 261–265
  definition, 261
  pattern, repetition, 268
Radioactive elements
  decay, 261f, 262
    rates, 269t
  energy loss, 261f
  usage, 275
Radioactive isotopes
  decay rate, 261
  naming, 259f
Radioactive tracer, definition, 264
Radioactive waste, 275
Radioactivity
  characteristic, 258
  environment test, 243
Radiocarbon (C-14) dating, example, 170
Radiochemistry, 258
Radioimmunology, 274
Radionuclides, 273
Radiopharmaceuticals, 273–274
Radiotherapy (radiation therapy), 274
Radon-222, 259
  decay, example, 264
Rare earth metals, 46
Rate equations, 186–188
Rayon (processed cellulose), 230
Reactants
  concentration, increase, 189
  enthalpy, 179
  joules per mole, 184
  products
    blockage, 184
    conversion, 178f, 185f
Reactions
  activation energy, reduction, 185
  equations, examples, 187
  equilibrium constant, calculation, 188
  heat, 179
  involvement, 183
  order, 186–188
    equations, 186t
  rates, 183
    equation, 186
  types, 261

Receptor, definition, 242
Reducing agents, 147–148
  definition, 147
Reduction, 144–145
  occurrence, 150
  oxidation, opposite reaction, 146f
Reduction-oxidation (redox), 144–145
  reactions
    balancing, 145–148
    knowledge, 155
Reference state, definition, 176
Rem, radiation measurement, 272
Representative elements (groups IA-VIIA), 41, 43
Resistance
  changes, measurement, 168
  measurement, 155
Resonance theory, 121–123
Reversible chemical reactions (change), Châtelier's
    principle (usage), 189
Ribonucleic acid (RNA), 225
Ring structures, 199–200
Rounding, 14–15
  example, 14–15
  rules, 14
Rowland, Sherman, 253
Rubber, resistance, 154
Rutherford, Ernest, 24, 41

S

Saccharide, 229
Salicylic acid, formula, 214f, 222f
Salt, solubility, 62
Saltpeter, molecular formula, 27
Saturated fats, 196, 233
Saturated molecules, 196
Saturation, 62
  description, 63
Scientific method, 6–7
  development, 6
  steps, 6
Scientific notation, 9–11
  example, 10t
Scintillation counter, 266–267
Secondary amines, 220
Second-order reaction equations, example, 187
Semiconductors, 47
  silicon, usage, 50
Semimetals, 50
Shell (electrons), 110
Siemens, definition, 155
Sievert (Sv), usage, 272
Sigma bonds, 122–123
Signature markers (molecular markers), 234
Significant figures, 11–13
  definition, 11
  rules, 11–12

Silicon
  abundance, 50
  characteristic, 50
  family, comparison, 195
Silver (Ag)
  atomic number, example, 129
  characteristic, 45–46
  conductor, 154
  ions
    doubling, 146
    reduction, example, 145
  naming, 260
  percentage, 49
  reducing agent, 147
  reduction, 145
  zinc, half-reactions, 146
Silver-mercury amalgam, 49
Simple cubic solid, 57f
Simple peptide, compound, 224
Single bonds, 123f
  formation, 196
  sigma bonds, 122
Single-stranded nucleic acids, hydrogen bonding, 224
Sink, definition, 242
Smith, Michael, 225
Smith, Robert Angus, 245
Soap, making, 232
Soddy, Frederick, 258
Sodium (Na⁺)
  cation, example, 135
  extraction, 144
Sodium bicarbonate (baking soda) (NaHCO₃), 138
  base, action, 161
Sodium chloride (NaCl)
  brine solution, usage, 100
  chemical formula, 27
  electrical magnetism, 58
  solution, 96
Sodium hydride (NaH), 139
Sodium hydroxide (NaOH)
  dissolving, example, 180–181
  hydrochloric acid, mixture (example), 162
  molecular weight, 99
  solute, moles (determination), 181
Soil
  testing, example, 260
  toxic contaminants, presence, 242t
Solidification, heat, 179–180
Solids, 55
  amorphous solids, 56–57
  characteristics, 56–57
  covalent solids, 59–60
  crystalline solids, 56–57
  density, 61
  heat, addition, 62
  individuality, 61t

Solids (*Cont.*):
  ionic solids, 58–59
  metallic solids, 58
  molecular solids, 57–58
  nonmetal form, 51f
  precipitate
    chemical reaction, 62
    formation, 96
  precipitation, 62–63
  properties, 60–63
  quiz, 74–75
  temperature, 61–62
Solids bonding strength, 59f
Solubility
  rules, 96–97
  writing, 96
Solute
  definition, 96
  moles, determination (example), 181
  relative amounts, 100
Solutions, 95
  characteristics, 96
  concentration, reduction, 104
  dilution, 103–104
  quiz, 106–107
  turbidity, 244
Solvation, 138
Solvent
  addition, 103
  definition, 96
Sørensen, Søren, 167
Specific gravity, 65
Specific heat, 180
Spherical nanocrystals, 47
Spin magnetism, 111–112
  example, 112
Spinoffs
  NASA list, 5t
  partnering, 4
Stability, island, 267
Standard heat of formation (ΔH°), 178–179
Standard state, 176–177
  definition, 177
Standard temperature and pressure (STP), 88, 176
  conversion, 89t
Starch, 230
  polysaccharide, 231
Stars, hydrogen supply, 166
Steel, conductor, 154
Stereoisomers, 204, 206
  structural isomers, contrast, 206
  study, 207f
Stern, Otto, 111
Steroids, 231, 233–234
Sterols, 232
Stock, Alfred, 140

Stock System, 140
Strawberry molecules, 217
Strong acids, 163
    weak conjugate bases, 164
Strontium ($^{90}_{28}$Sr), presence, 260
Strontium (^{90}Sr), half-life (example), 268
Structural formulas, 28–29
    definition, 28
    stereoisomers, contrast, 206
Structural isomers, 204, 205
S-type orbitals, 110
Styrofoam, 253
Subatomic particles
    components, 22
    neutrons, 26
    protons, 25–26
Subparticle arrangement, plum-pudding model
        (Thomson), 23–24
Subshells, 115–118
Substitution reactions, 208–209
Sucrose, 230
    chemical formula, 28
Sugar, 229
    alcohol conversion, 65
Sulfate ion (SO_4^{2-}), 139–140
Sulfide (S^{2-}), anion (example), 135
Sulfite (SO_3^{2-}), 139–140
Sulfur atom, oxygen atom (attachment), 139–140
Sulfur dioxide (SO_2), geometry, 124
Sulfuric acid (H_2SO_4), 167
    change, 245
Supernate, 62
Supernate, movement, 62f
Surface tension, 65–66
Surfactants, 216
Suspension, 62f
Swamp gas (methane), 251
Symbols, 35–36

T

Table sugar (sucrose), 230
Technetium
    naming, example, 260
    Tc-99, medical usage, 274
Temperature
    activation energy, relationship, 185–186
    impact, 61–62
    measure, 16
Tertiary amines, 220
    example, 220
Testosterone, 233
Tetrahedral electron pair geometry, 124
Thallium (Tl-201), radioactive tracer, 274
Theory, definition, 8

Therapeutics, creation, 238
Thermodynamics, 175, 176
    first law, 177–181
    importance, 189
    quiz, 190–191
    second law, 181–182
    third law, 182
Third-order reaction equations, example, 187
Thomson, J.J., 23–24, 41
Thorium, power supply, 275
3-ethyl-1-pentanol, formula, 195f
3-methylbutanamide, example, 221f
3-nitrobenzenol, formula, 222f
Thymine (T), 225
Tin, reducing agent, 147
Titanium, specifics, 39f
Titration, 169
Toluene, formula, 214f
Torr (SI standard pressure unit), 82
Torricelli, Evangelista, 82
Total dissolved solids (TDS), environment test, 243
Total energy, 177
Toxic contaminants, presence, 242t
Toxic metals, deposition, 246
trans-1,2-dichloroethene, formula, 206f
trans-1,4-dimethylcyclohexan, formula, 214f
trans-1,4-dimethylcyclohexane, formula, 195f
Transferases, 227
Transition elements, 43
Transition metals (group B elements), 41, 45–46
Transuranic elements, 40
    human creation, 263t
Transuranium elements, 262
    formation, 267
Tricarboxylic acid (TCA) cycle, 228
Triglycerides, 232
Trigonal bipyramidal geometry, 124
Trigonal pyramidal geometry, 124
Triisopropylamine ($C_9H_{21}N$), 220
Trimethylamine, example, 220
Triple-bonded carbon compounds, 198f
Triple-bonded hydrocarbons, 208
Triple bonds, 123f, 198
    formation, 199
Triple point
    definition, 71
    location, 70f
Trisaccharides, 229
Turbidity, 244
Turpentine, solvent, 96
2-butanol, formula, 207f
2-hexene, formula, 197f
2-methyl-2-butane, formula, 197f
2-methylpentane, formula, 195f
Two-step conversions, examples, 15–16

Tyndall, John, 97
Tyndall effect, 97

U

Universal solvent, 96
Unsaturated molecules, 196
Uracil, 225
Uranium
  atomic number, 262
  inner transition metals, 46
  naming, example, 260
  power supply, 275
  U-238, decay, 264, 267
Urea (CH_4ON_2), 221
  creation, 194
  formula, 195f, 221f
Urey, Harold, 258–259

V

Valence, 37
Valence bond theory, 119–120
  atoms, bonds (alternation), 121–122
  molecular orbital theory, contrast, 120–121
Valence electrons, 115
  example, 116
Valence shell electron-pair repulsion (VSEPR), 123
Vanadium, atomic number, 115
Van der Waals, Johannes, 90
Van der Waals' equation, 89–90
Vanillin ($C_8H_8O_3$), 199–200
  formula, 200f
Vaporization, 65, 67–68
  definition, 68
  dynamic equilibrium, 69f
  heat, 179
  rate, deceleration, 69
Vapor pressure, 92
Vinegar (acetum), 159
  hydrogen ion concentration, 168
Viscosity, 65, 66–67
  definition, 66
  example, 66–67
Vitamin A, 232
Vitamin D, 232, 233

Vitamin E, 232
Vitamin K, 232
Volcanic eruption, gas release, 249t
Volta, Alessandro, 150, 153
Voltage, electrical potential difference, 155
Voltaic pile (primitive battery), 150

W

Waage, Peter, 188
Water (H_2O)
  angle, 124
  creation, 162
  density, 61
  ethylene glycol (CH_2OHCH_2OH), mixture, 73
  geometry, 124
  honey, contrast, 67
  molecular formula, 27
  molecule, oxidation number (finding), 148
  molecules, polar molecules (attraction), 215
  oxygen solubility, 244
  pH, 168
  polar molecule, 133
  proton, loss, 137
  toxic contaminants, presence, 242t
Waxes, 232
Weak acids, 163, 164
Wet cells, 150–151
Wet electrochemical cells, simplicity, 162f
Wintergreen molecules, 217
Wohler, Friedrich, 194
Wood, resistance, 154
Work, potential energy/heat (relationship), 177

Z

Zinc
  environment test, 243
  oxidation, 144, 151
    equipment, 145
  reducing agent, 147
  silver, half-reactions, 146
Zinc-carbon cell, invention, 152
Zinc chloride ($ZnCl_2$), electrolyte, 152
Zinc sulfide (ZnS), electrical magnetism, 58